MAGNETISM AND ITS EFFECTS
ON THE LIVING SYSTEM

Books by Albert Roy Davis and Walter C. Rawls, Jr.

The Magnetic Blueprint of Life

The Rainbow in Your Hands

The Magnetic Effect

Magnetism and Its Effects on the Living System

MAGNETISM AND ITS EFFECTS ON THE LIVING SYSTEM

by
ALBERT ROY DAVIS
and
WALTER C. RAWLS, Jr.

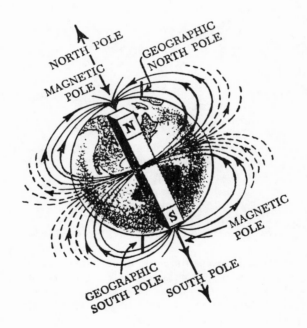

Acres U.S.A.   *Kansas City, Missouri*

First Printing, October, 1974
Second Printing, September, 1976
Third Printing, January, 1978
Fourth Printing, February, 1980
Fifth Printing, October, 1980
Sixth Printing, May, 1982
Seventh Printing, January, 1984
Eighth Printing, April, 1987
Ninth Printing, July, 1988
© 1974 by Albert Roy Davis and Walter C. Rawls, Jr.

Library of Congress Catalog Card Number: 74-84423

ISBN 0-911311-14-9

Printed in the United States of America

CONTENTS

PREFACE

This book was written to aid in the understanding of the Science of Magnetism, and its effects on mankind and all biological systems. We shall discuss how magnetism, that natural energy we find surrounding the earth, acts on living systems. Further, we shall in part discuss the effects magnetism and magnetic fields have on inorganic and organic materials, genes, cells, airs, and gases, as well as protein structures. To this day magnetism holds many unknowns and we shall present some of these answers to assist the reader, whether student, professor, doctor or physicist. We have tried to avoid complicated math and complex formulas in this edition, yet present new research findings that will afford a better comprehension of the effects of non-homogeneous magnetic energy. After many years of practical laboratory and clinical research encompassing the mechanics of practical and applied investigations in and with magnetism and magnetic fields, this research has developed new laws, theories, and new and practical understandings in all phases of the physics of magnetism and magnetic fields. This also includes electromagnetic fields generated by alternating voltages and currents, to direct currents and voltages, the solid state, and the standard metal magnets with their two poles of energy.

One of the breakthroughs discovered a number of years ago was that the two pole energies of any magnet are not homogeneous as to effects to any and/or all subjected materials, organic or inorganic in nature.

In the past few years your authors have continuously urged those in influential and senior positions with the government and scientific departments to reexamine the accepted theories and concepts of magnetism. However, our efforts have been unsuccessful and at this time many researchers at all levels of scientific research refuse to acknowledge or utilize this new discovery for guidance in their work.

The purpose of this book is to assist all of the men and women now engaged in magnetic research. We hope they will benefit from the research and new discoveries presented in this book, the extension of new uses for magnetism in all of its many fields and applications, even those not now thought possible, yet which within our foreseeable future will be understood and used by all nations of the world for the benefit of mankind.

April 28, 1974

Albert Roy Davis

Walter C. Rawls, Jr.

INTRODUCTION

Biomagnetism, biological use of magnetism, to aid and treat human and/or animal ills, is far older than the ancient science of acupuncture.

Acupuncture was brought into the open when the United States during 1973 reestablished relations with the People's Republic of China. This science is now under serious investigation here in the United States. After his rise to power in the People's Republic of China, Chairman Mao Tse-tung initiated the announced China Scientific Reconstruction Program, in which the Chinese are now making even further advancements and discoveries in this ancient science.

Although the Science of Biomagnetics predates acupuncture and holds many secrets man has yet to unfold and discover, we in the United States are not proceeding as we should to further develop this science.

Many of the writings and parchments were long ago destroyed by those ancients who researched and used this natural science. These parchments were destroyed, as were many others, to prevent them from falling into the hands of warlike people invading their lands and cities.

It is recorded that Dr. Hua To, one of China's most advanced medical men, born between A.D. 140 and 150, was the developer of acupuncture. China was then under the rule of Emperor I Tsung of the Tang Dynasty. This recorded evidence places a date that we can use as a reference to the ancient accepted use of the medical science of acupuncture.

To determine when magnetism or magnets were used in medicine, we refer to the works of the Greek physician Galen. We find that in Galen's ninth book of his writings entitled *De Sim-*

plicium Medicamentorum mention is made of a magnet's energies being used in purgative arrest and openings. This writing can be dated as far back as 200 B.C.—far older than the recorded science of acupuncture.

More recently, the works of Louis Pasteur tell of his research by placing a magnet next to fermenting fluids and wines and the marked rise in fermentation processes as a result. There are also the important writings, reports, books, and papers of such men as Von Reichenback and Walter Kilner and many others too numerous to mention. Hundreds of papers, books, and manuscripts have been written on magnetic fields and their effects on biological systems and man's environmental surroundings, animals and lower forms of creatures. In the investigation into any of these papers, there are exceptions to the rule, yet for the most part, these papers, books, and manuscripts indicate the failure to properly reproduce many of these experiments time after time with the same success, and many writings totally fail to show any effects of a magnet's poles and its energies on any forms of life, biological systems, serums, gases, or fluids.

Failure to properly develop the ancient science of magnetism is partially due to improper investigation of this important science by the scientific community. It is necessary to better understand magnetism before understanding and entering into physical or biochemical research and the practice of this research.

To assist the layman or student—with the forbearance of doctors, physicists, and scientists—we will present elementary data to acquaint the reader with the fundamentals of accepted magnetic theories, concepts, and principles which will provide a basic understanding of magnetism.

This book is not intended to cause any reflection on the work of individuals or groups who have written many books on magnetic effects, laws, physics, or principles without the little known and advanced concepts and discoveries that we have made during our research. We all must accept advancements to our knowledge of, and to all, sciences and scientific principles.

The authors give full and due credit to the years, even lifetimes, of the many men and women who have dedicated their lives to the study of the mechanics of magnetism in all of its many fields; also all of those whose work, papers, books, manuscripts, have been written to enlighten mankind by their findings, as by this effort on their part they have aided in the interest and progressive understanding of "What is Magnetism?"

In this book we intend to show the vast and far-reaching effects that may be obtained again and again, duplicating these tests and experiments anywhere at any time by any person equipped with the proper knowledge of how to proceed, making such presentations new scientific facts and not simply theories. All new discoveries, applications, and understandings presented in this book are by Albert Roy Davis or under his guidance and supervision from the Albert Roy Davis Scientific Laboratories in Green Cove Springs, Florida. This book is presented as a scientific paper as well as a book. There are excerpts in this book taken from patents that have been filed by your authors.

It is your authors' most sincere desire that the publication of these new concepts that are reproducible, the discussions and applications of these concepts, will advance the scientific knowledge of the understandings of applied magnetic fields and their effects.

Albert Roy Davis

Walter C. Rawls, Jr.

ACKNOWLEDGMENTS

Full credit and acknowledgment to all the dedicated men and women who throughout the years have researched the science of magnetic effects would require a book in itself, and then not be complete. We present here those researchers who primarily have assisted our work. Many of these scientists, in cooperation with our laboratory research, have duplicated in their laboratories our findings presented in this book. Some of these scientists were initiated into the science of magnetics through our efforts and then on their own made outstanding discoveries in their own right. In general, their combined assistance in duplicating new facts, disclosures, and experiments contributed to the establishment of magnetics as a basic important science to all sciences.

Any new breakthrough in any field of science, the discoveries that advance a science, must be reproducible at any time or location with consistent results by qualified persons. We have worked with other scientists in this country and other countries in this regard.

The fields of research mentioned in these acknowledgments are descriptive and not intended to be inclusive.

Dr. B. E. Roessling, University of Berlin. Magnetic wave propagation. Dr. Roessling's work in 1936-1937 brought about the first full understandings of the principle later to be discovered as the polarization of light waves—the now existing laser principle. His work also extended into the effects of magnetic wave emissions on and to living systems.

Dr. William Cambell, University of Cambridge. High frequency effects on living systems, 1935-1937. Dr. Cambell's work covered the physical effects of magnetic fields to rodents and small animals.

Dr. Yerkes, The Yerkes Primate Biological Laboratories, Orange Park, Florida, extension division of Yale University. Animal behavior in the presence of natural environmental magnetic biosphere, 1936-1937.

Dr. Blasengame, electrical stimulations of living systems and their effects, 1937 to 1938.

Dr. H. Bingenheimer, Germany, extension research. Applied electromagnetic energies to living systems to stimulate physical development.

Dr. N. S. Hanoka, University of Israel; resident, Harlington, Texas. Studies in natural sciences including food stimulation, product growth, seed treatments by magnetic fields, 1956-1971. Research into removal of toxic infections by magnetic fields. Assisted in our uncovering the reason why many wild animals can eat decayed and infected food with no ill effects due to the inner development of an antitoxin fluid. Aided in research into wound healing by magnetic field applications.

Dr. A. K. Bhattachura, West Bengal Clinic, India. 1959-1970, researched the effects of magnetic fields on living systems in the free clinics in India. Our joint research resulted in acknowledgment by Prime Minister Gandhi.

Dr. Edward Stadel, Applied Sciences, 1960-1967. Extension research, University of Oregon. The effects of magnetic fields on sugar beet seeds that resulted in a higher yield of natural sugars, and in tomato research the production of less acid end products, improved flavor, and better growth patterns, when using the proper pole of applied magnetic fields, supporting our findings of the dual effects of a magnet's energies.

Dr. George de la Warr, The Delawarr Laboratories, Oxford, England, 1966-1967. Research into intermodulation of electromagnetic frequencies in application to the human living system for the detection of biological effects, as a result of carrying other imposed electromagnetic frequencies together with a magnet's pole energies. Human and animal reactions to natural electromagnetic energies as found in the biosphere. Mrs. de la Warr was an active participant in this research.

Dr. Robert J. Morgan and Mrs. Hope Morgan, Delaware Clinic. Dr. Morgan, supported by Mrs. Morgan's untiring efforts, acted to reproduce research of the separate magnetic poles on inner ear defects, drainages, nerve responses, and forms of ear and biological infections. Further research toward the reduction and arrest of overactive acid conditions and the arrest and control of excessive acids produced by the digestive system of the living system. These were proved to be supporting on the results of the proper application of only one of the two poles' energies of a solid state magnet.

Mr. George Meek, Research Scientist and Engineer in Thermodynamics and Applied Sciences. Research into the effects of water treated with magnetic energies. Both he and Mrs. Meek made possible the field of researching the possible beneficial effects of a single pole's energies in the arrest and control of stages of cataracts and glaucoma. Experimental research on eyes of living systems affected with these conditions was outstanding both in efforts and results. Mr. and Mrs. Meek have during the past few years toured nearly all countries of the world investigating many new concepts in natural and biological sciences.

Dr. Ralph U. Sierra, Director, Puerto Rico Scientific Research Laboratories, Rio Piedras, Puerto Rico. Dr. Sierra has reproduced over 100 actual experiments that afford us scientific supporting evidence that the magnetic energy of each pole of the magnet is completely and totally different in its effects on the living system. Dr. Sierra's work and international lectures have provided many researchers with the results of our combined work into magnetic energy effects. His work in agriculture, medicine, medical, clinical and animal research has supported our work. He is one of the most active researchers in biomagnetics in the world today. His work, research, and lectures are known in many South American nations, as well as the University of Puerto Rico. Dr. Sierra has assisted in the duplication of our research findings now for some five years. He has afforded us with certain major discoveries that he has made in his laboratory in Puerto Rico in addition to his reproduction of our research developments and experiments here in Florida. His work has been shown in many nations of the world and he has given freely of his time and efforts to enlighten scientists who have visited his laboratory from these nations of the great values that this natural science of biomagnetics offers all mankind.

Dr. W. D. Chesney, Janesville, Wisconsin. Dr. Chesney has worked on development of many photosynthetic organic chemicals, the first fluorescent light and many other firsts in the field of applied science. He has assisted us in the research of applied biomagnetic energies and helped call this new science to the attention of the medical community.

Dr. L. Thornton Owen, Jr., Director, The Owen Clinic. Dr. Owen has undertaken research to duplicate our research in the application and promising results of arresting many complaints of the living systems. His work has been of great support and assistance in our research.

Dr. Harold H. E. Brownlee, Oshawa, Canada. Dr. and Mrs. Brownlee at their clinic in Oshawa have aided in the duplication of our research findings of the two effects the two poles of a magnet's energies have on the living system, each being totally different in nature and effects. Untiringly they have assisted us in proving that the two poles' energies of the magnet can and will effect a definite and scientific reaction when these energies are correctly applied to the human living system suffering from a complaint. Further assistance in nerve reactions seen or measured with suitable electronic instruments that indicate the location of these many complaints in diagnosing clinical investigations by the means of applied magnetic pole energies. The discoveries that have resulted from this research finding are now under full investigation in colleges in Canada. Dr. Brownlee's research into biomagnetics has been outstanding and valuable in the greater understandings of magnetic effects to the living system. Dr. Brownlee's work and untiring efforts to lecture and teach this valuable science to the medical and associated sciences in Canada have been received with great interest.

Dr. Ruth Wenrich Emerson. Dr. Emerson's work in the reproduction of our findings and research has been outstanding and rewarding in every respect. She is dedicated to the study of natural arts and sciences and performed an important part in opening new doors of research into biomagnetics in that magnetism, as that of a magnet's energies, is very similar if not exactly the same as the earth's magnetic fields, assisting us in this science as a natural science.

Dr. Frederick Doughty Beck. Dr. Beck's interest in the development of natural sciences, directed toward the improvement and development of a better understanding of natural methods or means of relieving suffering, prompted him to undertake to reproduce many of our experiments to establish this new science. Dr. Beck's work in kidney complaints and arrests has been outstanding, his own personal work supporting the singular effects of magnetic energies when applied to many complaints of the kidneys and other ailments. Also in magnetic effects on blood pressure.

Dr. Stanley Hall, Queensland, Australia. Dr. Hall's research and life's work has been dedicated to the natural arts and sciences. Dr. Hall's efforts in the reproduction of the effects to the living system by the singular effects of each pole of a magnetic to the living system have been of the upmost importance in supporting our work.

Dr. E. W. Hidson, London, England. Dr. Hidson, dedicated to medicine and allied sciences, has undertaken to prove a number of our findings and is now proceeding to explore still other new discoveries for the reason of duplication of findings to further support our research into the values of biomagnetics.

Dr. Leslie O. Korth, London, England. Dr. Korth's research into biomagnetics in his own right and in exploring and duplicating our findings has obtained some outstanding and remarkable scientific results on the effects of the two single and separate magnetic effects of the poles of a magnet in applying magnetic energy to living systems. He has been responsible for releasing information of our research work through such official journals as the *British Naturopathic and Osteopathic Journal,* 1973-1974.

Dr. George Walters, Florida. Dr. Walters' work on the effects of biomagnetics on the living system has been outstanding and he has undertaken to discuss and introduce this science to many scientific-minded men and women researchers throughout the United States. His aim is to make known this science to those dedicated to research and investigation of natural sciences, as has Mrs. Walters.

Mr. and Mrs. Lawrence Nelson. Mr. and Mrs. Nelson have devoted their work to presenting educational materials and research data to all interested in the natural sciences. They have also supplied much information and actual research findings in many fields of this science. Their work has been untiring in the exploring of the applied sciences including research into biomagnetics.

Dr. Leonard J. Allan, Margate, Kent, England, Osteopathic Clinic. Dr. Allan is the author of a number of books on natural sciences and diagnostic methods and systems in health care. He has assisted us in many ways to come to know and understand the effects of biomagnetics in many fields of biological research.

Professor Bessie O'Connor, Midnapore, Alberta, Canada. Professor O'Connor, an outsanding educator in Canada, has devoted her life to science, teaching, and applied research into the natural arts and sciences of the earth. She co-authored a book entitled *Magnetically Yours,* in which she presents a scientific look at the natural laws of magnetism in the study of plants, animals, and man in their magnetic biosphere. Professor O'Connor has assisted us in many research investigations that have supported and duplicated the work outlined in this book.

Dr. D. N. Khushalani, Rehmatbai Vadnagarwala General Hospital, Calcutta, India. Dr. Khushalani has researched biomagnetics with us for a number of years and is an outstanding medical and natural science teacher, investigator and researcher.

Dr. Earl W. Conroy, Kaita, New Zealand. Dr. Conroy's willingness to undertake and fully investigate sciences related to aiding health and locating new methods of combatting disease has been both outstanding and rewarding to our work. His help in researching the two singular effects of magnetic energies has been outstanding in every respect.

Dr. Yoshio Seki, Tokyo, Japan. Dr. Seki undertook the study of biomagnetics three years ago and has formed a new research program in Japan, far exceeding any work now being undertaken in the effects of the two singular pole energies of the solid state magnet to biological systems in Japan. While many doctors and scientists have visited our laboratory from Japan, Dr. Seki stands out as having the greatest potential of presenting this new scientific breakthrough in Japan and in other nations in Asia and Europe. He undertook this research under the most difficult conditions and has, as the result of his dedicated research and allied work, opened the doors to the further exploring of this vitally important science in Japan and now in other nations of the world. Dr. Seki has duplicated much of our work. Dr. Seki has undertaken to establish an international educational trade and post of scientific material exchange in many nations of the world.

Dr. Victor Beasley, North Carolina and Guyana, South America. Dr. Beasley is an outstanding scientist who has investigated most of the natural arts and sciences we find existing today. In one of his papers Dr. Beasley presents an outstanding review of scientific work now going on in most nations of the world, including the study of ancient beliefs, ancient medical sciences, man and his behavior, and also parapsychology investigations. Dr. Beasley has reproduced many of the basic discoveries we have made and has been of the greatest importance and value in this applied research.

Mr. Joseph F. Goodavage. Author of *Man, the Biomagnetic Animal, The Fabulous New Science of Biomagnetic Healing,* and other books and articles. Mr. Goodavage has assisted us in making contacts, in meeting and working with a number of fine scientists, researchers and investigators in the field of natural and applied sciences.

Dr. R. H. Gordon. Dr. Gordon, author of *Basic Studies on Monopolarity* a scientific discussion and review of magnetic effects on biological systems. He has obtained some remarkable discoveries as to magnetic effects. His work with his very talented sister throughout the years has resulted in obtaining patents on magnetic instruments now under research in a number of countries around the world. Our laboratories researched with Dr. Gordon in the field of photocolormetric investigations of visual studies of magnetic fields and developed a means of detection for visual display of magnetic fields in color for electron emission studies.

Mr. Clifford E. Swanson. Mr. Swanson is a publisher and one who through his untiring efforts after retiring from the publishing field has dedicated much time and effort to researching the effects of magnetic fields. He has been responsible for contacting and making possible many meetings with those interested in furthering their effective investigations into the science of magnetics.

Mr. McDonald Newkirk. Mr. Newkirk has over the past years shown a great interest in this science. He is known in India and in New York circles of research into natural sciences. He has made it possible for us to establish many new avenues of communications with a number of fine scientific men and women in New York and cities in other nations of the world.

Dr. Bernard Jensen. One of the developers of the science of "Iris-ology"; author and foremost authority in many discoveries into the investigations of the iris. Dr. Jensen has filmed some of our work. He hopes soon to present a film on his research into the environmental attitudes of man. He has been the guest of rulers of many countries. In his travels he has investigated the above-the-average life span of the people of many nations, including the people of Hunza and others. He has investigated the possibilities that where the earth's magnetic fields are the greatest, the highest, man's life may be affected in many ways. One effect very well could be the longer life span that certain countries' people show as the overall averages of normal to extended life cycles. Hunza has long been considered the long sought Shangri-la, as this nation which is located in the beautiful valley in the heart of the Himalayas and its people have shown the marked extension of man's normal life span. Dr. Jensen's films and books will present for the first time little-known facts regarding these subjects. Dr. Jensen has assisted us and we hope we have in part assisted his research, as we have found that life in years can be extended 40 percent or more with increased magnetic surroundings researching with many forms of blood-circulating animals and rodents, then why not mankind?

Dr. Marcus McCausland, London, England. Dr. McCausland has assisted in many ways in the introduction of our work and has aided in establishing many worthy contacts for us within the scientific community of researchers in England. He is an outstanding student of the arts and sciences and one whose help has proven of extreme value to our work and research here in the science of applied biomagnetics.

Mr. Chester Hurlbut. For many years Mr. Hurlbut has actively participated in continuing this field of research. Mr. Hurlbut comes from a long line of medical specialists and had it not been for circumstances would have followed in their footsteps. His untiring efforts held together for long periods of time the continuance of this work and research. No man would ever hope to have a finer friend. His outstanding work and research have shortened the time it has been necessary to spend to prove the effects of the singular yet totally different effects to biological systems of the two separate pole energies of all magnets. Mrs. Hurlbut has been an inspiration in his dedication and work in the many fields of research he has entered and undertaken. As a co-worker and advisor he has more than demonstrated the meaning of the word friendship.

Mr. and Mrs. Donald Larson and their sons, Roger, Donald and James, and their daughter Crystal, for many years of assistance and dedication in our research work.

Mrs. E. J. Leonard. Mrs. Leonard has provided us with technical and editorial assistance. Her active part in preparing the contents and layout of this book, her assistance and dedication to details, scientific projections, and presentations have been of valuable assistance to our efforts.

MAGNETISM AND ITS EFFECTS
ON THE LIVING SYSTEM

Chapter One

UNDERSTANDING MAGNETISM

LEGENDS AND RECORDED HISTORY

According to ancient legends a shepherd named Magnes was tending his flocks. Here the legends vary greatly. It is told that his staff made of iron was pulled by an unseen force toward a large rock where it was held and resisted the boy's efforts to free it from the surface of the rock. This rock mineral became known as The Magnes Stone. Today we call this magnetic natural material Lodestone. From the young shepherd's name, Magnes, we have magnet and magnetism—an unseen, untouchable energy that is the basis for the development of electricity as we know it today that furnishes the power for our lights, radios and television sets.

Since that time in ancient history scientists have probed this invisible force of nature that produced the first magnet, Lodestone.

To show the length of time this study and its legends have been known to man let us quote the following: "A.D. 597, St. Augustine came to Britain at the insistence of Pope Gregory I and as he viewed the magnets attracting one to the other and when they were reversed, the magnets opposed each other, with no hand touching either magnet, he stated out loud, 'When I first saw it I was thunderstruck.'" The magnetism of the shepherd Magnes has presented a great scientific question over many past and present centuries. To this day the true nature of magnetism is still far from being understood. Outstanding space researchers and world-renowned scientists are not applying the true nature and understanding that this important science has for application not only for magnetism itself but also for the application in the other sciences known to man.

Lodestone is magnetic iron ore or iron mineral ore; it is in part

3

the composition of the lava, the molten hot flowing lava of a volcano. As this hot lava flows up, then down the sides of a volcano, it slowly cools, and as it cools the earth's magnetism, the magnetic fields that flow from one pole of the earth to the other pole, passes through the lava and impresses on the lava these fields of magnetism. When the molten lava is cold it has accepted, stored and has in itself that amount of energy that existed on the earth at the time the mineral rock was formed.

THE MAGNETIC COMPUTER

The fields of magnetism stored in the rock have been used as a data computer that has told us a great deal about the history of the earth and its biosphere (biological atmosphere). Scientists utilizing their knowledge of the earth's gravitational magnetism during the history of the earth can more accurately date the evolution of fish and animal species. Science has pondered what effect magnetism, existing on earth in its fields of force and energy, had directly on evolution, the genes, size, life span, development of these species.

It is now accepted that fields of magnetism, strengths and weaknesses, have themselves not been constant but changed during the earth's history.

The initial important scientific work in this regard was by Dr. Normal Prime with the U.S. Geological Department. Also, the French physicist Dr. Bernard Brunhes in 1906 undertook to examine volcanic materials taken from the sides of many great volcanoes. He went deep into the sides, removing cores drilled from the volcanoes. Dr. Brunhes discovered that the lines of force in the removed cores of the magnetic rock changed directions in relation to the north and south poles of the earth's magnetic fields, by establishing the depth and related computed time with the lines of force or direction existing. After considerable testing and re-examinations this presented the fact that the earth's magnetic poles had reversed a number of times over millions of years in the earth's history.

Here Dr. Prime and his survey party, by means of atomic carbon dating, determined when these magnetic pole reversals took place. Later, oceanographic scientists, in dredging samples of shell and bone structure from the bottom of the seas and using atomic carbon dating, were able to approximate more accurately when certain types of fish and mammals different from those now present in the oceans and seas existed on the earth. The same concept was used on the earth's surface, taking the remains of fossils, mammals and giant animals. Computing their period remains on earth to the levels of magnetic energy then existing on earth we can see that magnetic energy, amounts, and the magnetic pole relationship at that approximate time to many types of animals, fish, mammals, from beginning to end of their existence on earth and the beginning of new strains of animals, plants, fish, mammals. In many respects this allows us a magnetic time computer to obtain, understand, and gain knowledge of the earth's history and its changes, many in part, if not all, related to the changes of the earth's own natural magnetic fields.

PRESENT USES AND APPLICATIONS

It is not difficult for man to make a metal magnet of a solid state. You may take a piece of iron or steel and place it in a winding of insulated wire, or wind a number of turns of insulated wire around a nail. Connect to a good storage battery for about 5 seconds, which allows the battery's voltage to flow through the coil and impart to the steel, iron or nail the lines of magnetic energy that the coil produces when connected to the battery. The result is a magnetized material, a magnet.

Commercial and industrial magnets used in biomagnetic medical and biological research vary in intensity. The measurement of magnetism is termed a gauss. A gauss is a unit of magnetism as a volt is the unit of measurable voltage and as an ampere is the unit by which current is measured. The gauss is one unit of measurement in the most elementary manner of units in the measurement of magnetic force.

In many accepted discussions of magnetism it is advised that the earth and a magnet are alike in nature in that lines of force transmit from the North (N) pole to the South (S) pole. Other text references advise the lines of force transmit from the South (S) pole to the North (N) pole. However, after many years of research into the physics, physical and applied research, and investigation into magnetic behavior, we must correct this impression. Practical examples and laboratory research conducted many hundreds of times show that the lines of force that travel between the poles of the earth or those of a magnet travel not in one major direction but in both directions at the same time. This and other findings we will present at length.

It is of the utmost importance that you understand how we arrived at identifying the poles of a magnet, as many present-day magnet manufacturers do not code or identify the poles correctly. The two poles of any magnet are the N pole and the S pole. As does the earth, a magnet also has its two poles. The simple means of identification of the two poles is to take a long straight bar or cylinder magnet and tie a string or thread at its center. Then tie the thread to a support that will allow the magnet freedom to swing freely, keeping it away from all metal objects. The magnet will turn and slow, then stop turning. The end of the magnet that is pointing to the N pole of the earth is "the S pole of the magnet." You may code it for identification with red fingernail polish or red paint. Many references are given to the north-seeking pole of a magnet. This would mean that, since dissimilar poles attract and similar poles repel, the end seeking the N pole of the earth's magnetic pole is the S pole of the magnet.

The making of any magnet is the aligning of the atoms of the material. When you place a nail, steel or iron rod or bar, or other materials that are magnetically sensitive in a coil of current you align the atoms so they spin. Their electron spin is all in one direction. Therefore, the strength in gauss units of magnetism a magnet can be made to have, depends on the number of atomic shells within that material that contains a varied number of atoms that can then be magnetized or polarized. Now, while the electron spin of the atoms is aligned in one direction, each resulting pole of any magnet gives off energies that spin in opposite directions. The S pole spin is always to the right, while the N pole electronic field spins to the left.

Magnets today, with the advancement of magnetic material research, are made of plastic compounds mixed with certain magnetically acceptable materials. There are also magnets made from many kinds of minerals, noted and referred to as rare earth magnets. Therefore, it is now possible to make nonmetallic magnets and magnets that are flexible in the form of magnetic ribbon, tape, etc.

The use of a flat piece of paper with iron fillings placed on its top and the bringing up under that paper a magnet to show the magnet's lines of force is incorrect and should not be used in textbooks of many types to educate students, because each fine particle of the steel or iron fillings when placed in the field of the magnet under the paper becomes a miniature magnet in itself; thus the total picture is incorrect and misleading. As each miniature magnet then attracts and repels, the picture is distorted to present a mistaken concept.

Modern educational concepts and teachings of the principles of magnetism are to a great degree incorrect. Textbooks today still present the energies coming from a magnet as leaving the North or the South pole of the magnet to circle the full length, or between the poles, if the magnet is shaped like a horseshoe, and reenter the magnet at the S pole or in some texts the opposite. Again, here we have two errors that should be updated in all textbooks that teach this very important science to students and new scientists.

The first of the two errors is the belief that the energies always leave one pole of the magnet and travel to the other pole. This we have researched with practical and scientific studies, and the findings are that the energies leave the S pole and then flow to the N pole, the gravity and/or magnetic vortexes (circling cable-form-like energies) actually flow from the S pole to the N pole showing energies, waves of gravity held motion, in that direction. However, the vortex (circles) of magnet energies travel in both directions, S to N and N to S. This we proved in the magnetic flux (lines of force energies) as seen in the movement of the hydrogen bubbles in the two poles magnetic field movements. This test consists of a microscope slide, a few drops of diluted sulfuric acid, a medium power microscope, placing a magnet at each end of the slide, the diluted acid touching each magnet. Microscopic viewing after a few minutes allows one to see the energies of the two pole effects and the two directional movements of the sulfuric acid hydrogen bubble movement. A similar test using whole blood shows the spin of the red blood cells when placed in the field of a magnet. Taking whole blood, then spinning off the fluids and plasmas, leaving the red cells, presents a very remarkable piece of evidence as to the effects of magnetism on life fluids. Take some of the resultant red blood and place on a microscope slide in a good powered microscope, focus, bring up under the slide's bottom one end of a magnet. Note that the red blood cells all spin around the same direction. This is polarization of the red blood cells. Reversing the pole of the magnet to the blood sample reverses the spin or polarity.

We will show later how this enforces the red blood cells as to the electrical cell effect and the organic iron complex effects of the blood in part.

The first error in present teachings in part, of magnetism, as we have presented, is the failure to teach the opposite direction of flow of these energies.

The second error that is taught is that magnetic energies flow in a semicircle from one pole to the other pole. Again, this is incorrect. The simple test to support this incorrectness is to take a three- to six-inch bar or cylinder magnet and place it on a wood or plastic table, any base material that is not magnetic. Next, take a straight pin and, holding it between the thumb and index finger, place it at one end of the magnet. Moving the pin very slowly the length of the magnet, maintaining the slight upward pull, yet keeping the pin in contact with the magnet, at the exact or almost exact center the length of the magnet you will find one fractional place at that center where there is NO PULL. Therefore, no measurable amount of magnetism exists at the direct center of the magnet. This experiment will apply to all magnets in principle. In fact, the magnetic vortex (cables of circular energies) when leaving the S pole of the magnet travels to the center of the magnet and changes its degree of rotation by 180 degrees, then spinning in the opposite direction, continues on to reenter the magnet at the N pole. When the energy leaves the S pole of the magnet its vortex is spinning to the right. On reaching the center of the magnet the energy changes from positive to negative by a phase change of 180 degrees. Then, at this point, the vortex is spinning to the left. The left-hand spin is negative in energy to the right-hand spin which is positive. The lines of force are then divided into two different pole energies, north being negative in respect to the south being positive in electrical biological and potential force effects.

This completes the two errors in the presentation of magnetic principles as now taught in textbooks and many accredited schools and still followed by many scientists and research persons in the scientific world.

A further discovery from many experiments and years of research is that each pole of a magnet has a completely different effect to all subject material to which they are applied or come into contact. The common belief that the energies flowing between the two poles of any magnet are homogeneous, the same, is totally in error and is incorrect and has led for hundreds of years, if not thousands of years, many researchers, scientists in the wrong direction as to effects they obtain when exposing living systems, biological matter, to the fields of a magnet or magnetism. The present books written in Russia, Japan, and by members of colleges, universities and government-sponsored biological researchers fail to accept these discoveries as fact. Not until the birth of the Space Age have government scientists been able to test, see, deep in space, that the division of magnetic energies exist. In the Albert Roy Davis Scientific Laboratory the correct divisions of a magnet's energy were first discovered in 1936. This discovery, and many of the applications from this discovery, were brought to the attention of scientists in the government and to a number of other well-known and respected members of the scientific community. The results have not been satisfactory to the advancement of this science. There is ample indication that government-sponsored scientific investigations as well as the scientific community in general do not inquire in the proper manner into new developments of magnetism. Scientists suffer with the fault of many in that they find it difficult to learn new principles and accept changes that are contrary to their textbook teachings and their accepted theories in the research of magnetism and related sciences.

However, if we consider the thousands of years acupuncture has been in use serving the people of Asia, and mainly China, then we might see how new concepts not native to certain countries and serious investigation into these new concepts may have a very slow start regardless of their importance to man.

Biomagnetics is in this class of delayed action on the part of all nations not now investigating its potentials to serve all mankind with new and important discoveries. The Russian scientists have since World War II made outstanding and highly progressive steps in the new and unknown applications of magnetism in all of its many fields and possibilities. Yet scientists in America and other countries are slow even to consider serious research into magnetism. We hope this situation will soon change.

There are many forms and types of magnets. There are besides the standard metal magnet many forms of electromagnets. These are made of soft iron cores with many windings of wire over the core, each layer or winding insulated from the other. These electromagnets are considered just that and nothing more. However, electromagnets differ greatly from the solid state magnets, metal magnets or composition magnets in that they have a different effect, as has been shown in many research applications using both types of the same power in gauss units of magnetic energy.

In this book we shall endeavor to show these differences and explain in part the effect phase differences as well as the biological effect differences. We plan to show in a later book the more advanced discoveries concerning research findings from imposing other energies on the existing magnet poles in securing greater effects to many forms and types of living systems. Further writings are also planned concerning the more advanced effects on chemicals by the use of certain magnetic field forces. Here we point out that there must be a better scientific climate toward magnetism than now exists for these detailed discoveries to take their proper place in science. In general, laboratory findings show that chemicals change weight under certain magnetic field forces. The gravitational pull is altered, therefore, the weight of material, fluids, airs, and gases. This laboratory work also encompasses the tissues, chemicals, and fluids of the human and animal systems. Part of these findings are discussed in a later chapter.

There have been some noteworthy and rapid breakthroughs recently in magnetics and biomagnetics. An example is the manufacture of electricity without generators or turbines. Other important work has been accomplished in medicine, chemistry and physics. The discoveries presented in this book should assist the scientific-minded person, the student, doctors, scientists, researchers, toward a new and greater understanding of how to better develop this great and important science, and to this end we intend to open our research files and release certain new data that is in need for further research and development into the true nature of magnetism and what it can offer for mankind.

Chapter Two

DISCOVERIES MADE INTERNATIONALLY SUPPORTING INTERNATIONAL WORK IN BIOMAGNETICS

Today no research of importance "should be undertaken" in an isolated atmosphere, because, to avoid duplication of work, contact should be maintained with the rest of the world and one's country as to the many fields of research and developments undertaken by scientists and investigators. It is for this reason we advise you of the research work and progress that we are aware of in magnetics and that is now being carried on in many nations of the world today. We base these statements in part on direct visits from many scientists from many nations who have come to our laboratory to see and study our research into biomagnetics that we have been conducting for many years.

Our findings as to present-day research are based on actual and factual discussions with men and women from many nations and their understanding and knowledge of the work being carried on throughout the world today. We have also carefully investigated papers, books, manuscripts, scientific publications, releases, made by other nations through their own publications. This is also research, and it continues to take a great deal of time to check into all of these reports. However, the results have furnished information that has saved much duplication of effort and research in our work. It gives a better understanding also how far advanced we are in certain fields of development in this science. In this regard, the results of these studies have been very rewarding.

Should you ask, "Who is leading in biomagnetic research, what scientific group or scientist, what nation?" Based on the information released by publications (1960-1974) reflecting the research work-

11

ing being carried on in the Soviet Union, we find that Russia and its scientists have since World War II made remarkable progress in this work. The United States has, through its advanced space program, made a number of important discoveries and developments dealing with low fields of magnetism on man's environmental biosphere (biological atmosphere) in space, weightlessness, and the moon's low gravity and magnetic fields. Still little is made known by the U.S. on any work in the research toward aiding mankind by the use and applications of magnetic energies to man's biological system.

England for a number of years has lowered the barriers to research in many fields of medical work. Research scientists in England have also done some work in magnetics but not as much as one would have hoped considering the importance of this science. In France, the research work is also lagging in keeping up even to a small degree with the rest of the nations now investigating and researching magnetic energies. However, Russian and French scientists have now started to exchange information and work together in some fields we are aware of and these are in marine biology, the physical sciences, and biological sciences in general. The Russian and French marine and oceanographic scientists have discovered that fish have a built-in magnetic computer allowing them to orient themselves to the earth's magnetic poles and the earth's magnetic fields. This is a navigational system of nature superior to many of our technology achievements.

One interesting experiment the Russian-French team undertook was to transport a number of fish from France to Kaliningrad, Russia, a considerable distance from France. They were placed into a path-finding aquarium system. It was shown that the fish oriented themselves and would swim in directions to avoid the earth's magnetic meridian (0 to 180 degrees). Birds also have a magnetic compass and use it for their navigational flight systems and directions in storms, bad weather, as well as normal flying. Birds, crickets, bugs, beetles, it was found, when landing after flight movements, come in for a landing from either a north-south direction or east-west orientation, using the north-south direction of the earth's magnetic pole directional flow path as a guide.

Scientists from France, Germany, England and the United States, and other countries are aware of the noticeable affects of attaching a magnet, as an example, to a bird where the navigational system refuses to work. Many papers by world scientists have been written on close relative subjects to magnetism so we cannot give all the credit here to the Russian or French scientists. It is regrettable that many leading French scientists have refused to adopt magnetic effects to aid mankind since in their own country some very excellent work has been conducted in biomagnetics and magnetic sciences on biological systems, including man, that show it can be used as a great new tool in the field of applied medicine and the medical arts and sciences.

Japan's scientists and investigators fail to undertake serious and practical research into the effects to the biological system. Their work is so divided between the varied sciences as to make it difficult to understand what direction they are taking other than commercial developments.

We have had visitors from Japan, medical doctors and scientists, interested in learning more about magnetic effects who consider their own work the most important. This is natural, yet it fails to show the proper interest for one of the greatest sciences man has yet to uncover. In northern Japan, manufacturing and commercial interests have designed, developed, manufactured and presented to the world market instruments for the treatment of many of man's complaints. In investigating these instruments we find that little practical knowledge, medically or scientifically, is understood. This is not the kind of instrumental designing and offerings to the public that should be made without proper and detailed knowledge of the biological and medical scientific findings, as we have discovered in our laboratories.

However, we must remember that today the understandings in all respects as to what is, how does, acupuncture work still must be investigated to present the answers. Why and how scientifically does acupuncture work? The explanations are not complete. Comparing with many drugs that have been used for years in the world, no governmental agency can tell you how they actually work. While this is far from the proper approach scientifically, these matters now stand for all to question. How do they work? One of the reasons this book was written was to present certain facts that are new facts. Why and how does biomagnetics work? To aid in clearing misunderstandings, lack of basic knowledge, and further to introduce new discoveries that may further explain. We hope to show why biomagnetics will open the doors to new approaches in all fields of medicine and sciences.

In Canada we find that private research is outstanding in many fields of research. However, government is still not very interested. They are still to understand what this science holds for all mankind. At this time, through the efforts of a number of fine scientists, two universities have undertaken to research biomagnetics and its effects on man. We have had visitors from Canada, scientists who have made outstanding discoveries in biomagnetics, who cannot reach the government scientific community. This is equally true at this time in the United States. Many of our own scientists cannot get through the old and well-worn bureaucratic departmental roadblocks to show, explain, present many new sciences. This is not a new story as history records this very clearly throughout the ages of man's struggles to advance the arts and sciences.

A number of Israeli scientists have communicated with us and they show a great interest in biomagnetic research and development, so there is hope here that some new discoveries may be forthcoming soon. We would like to see other nations in Asia and near Israel take an interest and get more involved in the investigation of this science.

India shows promise; yet, again, independent scientists are doing the greater part of the research work. The science of biomagnetics was introduced to India's clinics about 15 years ago by the Albert Roy Davis Research Laboratory and much good has resulted from dedication to the understanding of this science in India. However, little attention is given to this science by government agencies or government scientists or the nation's leaders. Again, we have found in all cases, the governments of all nations, except Russia, depend on their senior scientists delegated the responsibility of investigating new sciences. Scientists need to be educated to understand and accept new concepts, leaving the old and outdated modes behind them. This is not now the case.

With the assistance of Dr. A. K. Bhattacharya of West Bengal, India, we undertook to introduce biomagnetics as a humane science. After a number of years of work we co-authored and published a book in India entitled *Magnets and Magnetic Fields*, which was directed to biological use and understandings. Since publication in 1970, it has been presented in many nations of the world. However, its contents present limited concepts that need updating and should no longer be used as any degree of effective present-day research.

On May 7, 1971, we received a letter from the Prime Minister of India, Indira Gandhi, in reply to a personal letter advising her of the magnetic materials, information, direction and assistance to India over a number of years and with the assistance of Dr. Bhattacharya how we had introduced the research of biomagnetics as a humane science into that nation with the resultant publishing of the book *Magnets and Magnetic Fields*. Prime Minister Gandhi's reply was one of great interest and she pointed out the future was in the hands of the scientists and politicans of the world and the responsibilities for future generations in their hands to use wisely.

A copy of the book mentioned above was requested by the Smithsonian Institution, Washington, D.C., where it is now on display.

German scientists engaged in private research into biomagnetics are the only ones active in this field in Germany. Germany has always been a leader in new scientific developments; however, like other nations, Germans have lost their eagerness to explore the unknown. We have had visits from and active communications with German scientists. Nevertheless, they are far behind in this research program.

Again, we find no government interest in biomagnetics in South America, yet more work has been done in these countries by private researchers than in many Asian and European countries. One researcher in Puerto Rico has gained much attention for his dedication to this science, Dr. Ralph U. Sierra.

A number of South American scientists and an American in South America have done outstanding research in presenting the science of magnetic effects to their respective nations. Dr. Ralph U. Sierra of Rio Piedras, Puerto Rico, has accomplished much work and great interest in that country, as has Dr. Victor Beasley, the American in South America. These men have shown outstanding understanding of the biological effects of magnetic fields to the living systems and have devoted much time and effort to promote interest in this science.

There is no doubt that other work in the field of research and development is being carried on in other small and large countries today, yet it is for the great part unknown. We will direct our attention to that work and research now in progress in those nations allowing some of their work and research to be published.

Electromagnetic Fields and Life written by Dr. A. S. Pressman of Moscow, Russia, is one of 30 books written by Dr. Pressman covering a wide range of well-conducted research reports into magnetic, electromagnetic effects of magnetic waves, and encompassing many phases of electronics in medicine. His work covers and includes microwave effects to the biological system and hygenic evaluation of high frequency electromagnetic fields. However, no information on the effects of the two separate poles of a solid state magnet energy source can be found. Also, we find no reference to the splitting of the magnetic poles at the equatorial axis of the magnet and/or the independent pole effects. Therefore, it is our belief that this discovery made in 1936 by Dr. Albert Roy Davis is not known nor has it been investigated by Russia or any other nations of the world that are investigating biomagnetics. This discovery we made and have worked with for numerous years will, we feel strongly, promote the advancement of magnetic research and developments in medicine, chemistry, and biological physics, as well as applied physics of magnetism. As we have previously stated, the accepted concepts and laws of magnetism are that both poles of any magnet or electromagnet are homogeneous. This we will establish in this book is in error. This basic discovery should be introduced into the physics of magnetism as a new law, principle, and concept, with approaches of better applications and understandings of magnetism and its effects on all modes, systems, developments, and also its great value and importance in the fields of medicine and biological sciences.

As we continue to review the work that has been done and is now underway in the Soviet Union, let us remember that all comments made, tests, experiments disclosed, treatment of the human system, animals, etc., are conducted with both poles of an applied magnet, as they do not have the information we will release herein as to the two separate pole effects. They assume, as do most scientists, that the two poles of the magnet produce a homogeneous energy.

In certain Soviet releases references are made to very important research conducted in 1948 when Red Army specialists used magnets to reduce and relieve advanced leg pains after and/or before amputation. During World War II Russian doctors, also reported in some of the Soviet papers, used magnets to relieve pains from wounds suffered during engagements on the battlefields. They also refer to the application of powerful solid state metal magnets to speed and/or reduce the length of time for healing of wounds that nature normally requires to make such recoveries.

Canadian doctors during the world conferences on electrosleep and electroanesthesia held in Bulgaria, September, 1972, read their papers on the use of magnetic energies to speed wound healing. This then allied the prior work by the Russian scientists.

Russian scientists have heretofore established the term "magnetobiology" as a new and important practical working science. The Russian scientists have developed a form of magnet that can be attached to the wrists of patients, again using poles equal in strength and opposite in potential. At the Rostov Medical Institute these magnetic wrist-connected magnets are used to assist in the treatment of certain types of heart and nerve diseases.

The Russian scientists continue to report that when a magnet's fields are applied to blood there is a rise in the effects as to coagulation and have also noted profound effects and changes during blood transfusions.

At the Leningrad Military Medical Academy they have shown effects to water when water is subjected to magnetic fields. They further show in the application of a magnet's poles to a human system the lowering of certain types of high blood pressure conditions. In several Soviet press releases it has been stated in no uncertain terms that aspects of their research findings are classified due to military use. No doubt they have discovered that biomagnetics and magnetobiology can be directed to new concepts in military applications. The terms biomagnetics and magnetobiology mean the same yet are described in Russian works as semi-universal terms —words that have the same reference meanings. The Soviet scientists are at this time very interested in low magnetic field effects as to possibly making some outstanding discoveries in the probes now underway of Mars and Venus. It is interesting to note that where the biological subject is in an environmental attitude of low magnetic field for extended periods of time the results can be deadly. There are midranges of magnetic fields of certain strengths where the best or worst effects can be found. Above or below these strengths there could be little or no effects in evidence to the living system.

There is no question in anyone's mind that the Russian research and developments are outstanding in every respect. During 1972-1973 the Soviets sold to American industrial manufacturers patent rights, leases, in advance use of a magnetic magnet's energies as a better way to produce aluminum finished stock. Their work in the development of hydro-magneto-dynamics (HMD) again presented their advanced concepts and applications of the two poles of a magnet's energies. HMD is the passing of a super hot gas, in the beginning the gas was seeded with cesium, through the two poles of a magnet. After leaving the magnet's fields it forms D.C. electricity and is thus collected on two electrodes, one a positive collector and the other a negative collector. We then have the generation of electricity without boilers, turbines, or generators. At this time American transformer manufacturers are building the largest electromagnet ever made for a giant HMD generator which will surpass several large city power generator systems. This is but another development in the many new discoveries in magnetism. Our laboratory has a number of new developments that offer great promise in a number of fields. Magnetism is acting to slowly change man's concepts for designing power systems and also new concepts in all fields of industrial, chemical, and biological uses and applications all revolutionary to accepted scientific understandings.

It would require several volumes to properly cover the research work and developments of all the known discoveries of the past 10 years. Russian scientists have and are making great strides in the research and development of magnetics.

At this time certain Japanese firms are manufacturing a magnetic dual pole, magnet bracelet, and making claims we cannot, nor can the U.S. government investigative agencies accept, as little or no practical medical evidence is offered that would technically support such claims. However, we do know that when magnetic energies are correctly applied to any part of the body there is a reaction and this reaction may be used, if properly applied, for the aid and/or relief of many animals and possibly human disorders. The work in our laboratory has been confined to all types and kinds of large and small animals and bacteria strains. As a result we can refer to reactions of blood-circulating animals, those that are similar to the human system, and make direct comparisons as to possible effects that may result when applied to man. Many of our findings as a result of animal research have been duplicated by doctors and scientists in other countries with outstanding and rewarding results.

We have had lengthy conferences with a number of Canadian scientists and doctors engaged in new work in biomagnetics. One is Dr. Harold Brownlee located near Toronto, who has researched with us and in the development of a system for the detection of human and animal ills and diseases. This is founded on the reaction of the affected organ or segment of the body to the field of an applied magnet unit. The fields that are designed and applied then act to cause a physical body response to indicate the area where the condition exists. Its possibilities as an analytical tool are very promising. We understand several universities in Canada are soon to test Dr. Brownlee's and our development on human subjects.

Canadians engaged in biomagnetic research are increasing each month and year. Another dedicated researcher is Dr. Bessie O' Connor, who is undertaking to study magnetic effects on and to agriculture and biological tracings by magnetic fields as to genetic effects.

Many doctors and scientists connected with colleges and universities in Canada are working to improve seed germination by exposing seeds to magnetic fields. The results indicate a 10 percent or better plant yield of its products on harvesting. Seeds when exposed to a magnet's energies can and do show remarkable effects of this energy. We will in forthcoming chapters describe from our research and findings how and why this stimulation takes place, how it can be more effectively used and how improved results may be obtained in applying our discovery of the singular use of each pole of a magnet rather than the use of both poles at the same time as is now the procedure used in these types of experiments.

The living system, regardless of type, kind, size, or nature, is subject to the added life stimulation seeds are, and comparisons will be made later in this book. There is a limiting and controlling energy application also.

We have only briefly presented work now underway in the biomagnetic sciences in many nations of the world. We will discuss some of the countries we have not covered in later chapters. We will present our research, compare it to work being done by other countries, and show where advancements can be made by the use of what we have researched and discovered in the same fields.

While we have not named all the doctors, scientists or countries engaged in this work, we stand ready to acknowledge any and all persons, their work and their country's efforts in this field of scientific research.

Chapter Three

MEASUREMENT OF THE EARTH'S MAGNETIC FIELD—THE OLD AND THE NEW CONCEPTS—THE DIVISION OF THE POLES

On page 22 we present a graphic series of drawings to show what is found in textbooks for instruction of students in the accepted theory of magnetic energies as they surround the earth and also a magnet. This is the popular belief concerning the movement of magnetic energy around the earth and also a magnet.

On page 22 we present the updated concepts from our findings, initially made in 1936 as to the division of the two poles' energies, each separated one from the other and each having a different potential, value, in electronic magnetic currents. The south (S) pole is positive in respect to the north (N) pole, which is negative.

Referring to page 22 you will see that in the use of a straight bar or long cylinder magnet, the two poles can be used each separated one from the other, and only the pole you wish to work with is then applied for exposure of any system you may wish to apply it to.

The conventional horseshoe magnet is not suitable for use in the application of only the one pole's energies as the poles of the horseshoe magnet are too close together to allow isolation to the degree we can have, by the use of the straight type of magnet.

The strength of the earth's magnetism in gauss is now only approximately one-half of one gauss, a very low magnetic field when we compare what the earth's magnetic strength was many millions of years ago in relation to the core drilling results from the sides of the great volcanoes and the atomic dating employed. The earth's magnetism has been many hundred times higher than its present strength.

Earth's Magnetic Field

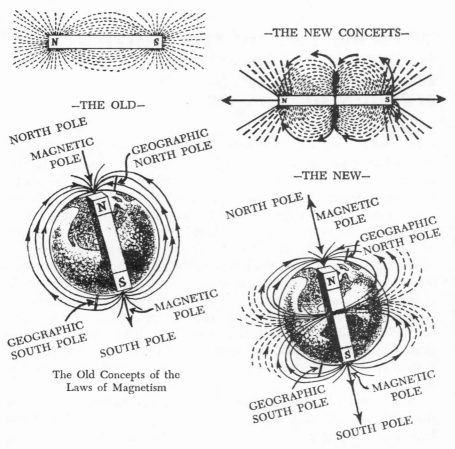

—THE OLD CONCEPTS—

—THE NEW CONCEPTS—

—THE OLD—

NORTH POLE
MAGNETIC POLE
GEOGRAPHIC NORTH POLE

—THE NEW—

NORTH POLE
MAGNETIC POLE
GEOGRAPHIC NORTH POLE

MAGNETIC POLE
GEOGRAPHIC SOUTH POLE
SOUTH POLE

The Old Concepts of the
Laws of Magnetism

GEOGRAPHIC SOUTH POLE
MAGNETIC POLE
SOUTH POLE

The New Concepts of the
Laws of Magnetism

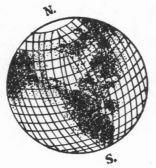

The direction of flow of the field of magnetism that surrounds the earth is shown on the drawing of the world, page 22, as traveling from the S pole to the N pole; see arrows indicating direction of flow. Many textbooks refer to the direction of flow from the two poles as from the N pole to the S pole. However, this is incorrect, and this new discovery can be supported by laboratory and space findings in the last few years.

The drawing shows a bar magnet having the conventional two poles. In the direct center of the magnet is the Bloch Wall, or the point of division of the circling vortex (spin) of electronic magnetic energies. The small arrows shown on the bar magnet indicate the direction of the spin of each pole's energies. The center of the magnet shows the phase change of the spins.

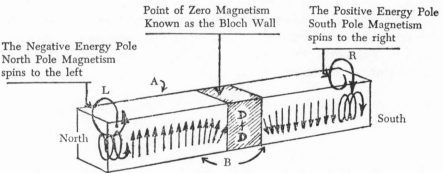

The north pole or negative spin is counterclockwise, or to the left. The south pole or positive spin is clockwise, or to the right. With the use of a straight bar or cylinder magnet we may then have access to the two separate forms of energy for our application of just that energy form and/or type. The illustration and discussion on this page is an outline of our initial 1936 discovery.

The biological effects of the application of the poles to biological fluids in a living system also show the path of travel to be from the S pole to the N pole, and this also includes all magnets as we have presented in the drawings herein. We have superimposed a magnet over the drawing of the earth on page 22 to show this flow. For the correct division of the poles refer to the drawing on page 22.

As mentioned earlier, the energies leaving each pole of a magnet form a line of almost straight energies that appear to travel great distances from the end of the magnet and appear then not to return to the magnet. This we believe accounts for the loss of energy of any magnet that does not have a keeper. A keeper is a simple shorting bar of steel or iron placed across the poles of the magnet to keep the magnet's energies circling within the structure of the magnet and avoid the loss that would happen should we not apply the keeper bar. The lines of force described as straight lines of force that leave the end of the magnet and do not return to the magnet end in no way alter the circling force field that remains the main attracting and repelling forces as presented in the drawings on page 22.

We are closing Chapter Three with the discussion and drawings presented only to better explain magnetism, the rate of flow direction, the two poles, the division of the two poles, and the potential force as to the electronic charge potential of each pole. This allows a reference to scientists who may not have continued into the physics of electrodynamics. Students and instructors should attempt to upgrade and correct any older text materials that differ from today's concepts and understandings of the basic science of magnetism as presented herein.

Chapter Four

THE EFFECTS OF THE TWO POLES ON THE LIVING SYSTEM—THE DISCOVERY OF THE TWO EFFECTS

T he primary discovery that the two poles of a magnet act to change and alter biological systems in two completely different ways was made by Dr. Albert Roy Davis in 1936. This discovery was quite by accident, and the noticeable effects on two cardboard containers of earthworms led to undertaking a serious study of this change to the biological systems.

The accidental discovery was made in a small home laboratory built upon leaving grade school and prior to attending the University of Florida. A large horseshoe magnet was on a wooden work bench near work on a small electronic oscillator used in the old superhetrodyne radio circuits in early days of radio. The afternoon had been planned for fishing, and three cardboard containers of earthworms were on the workbench. The earthworms in the cardboard containers were in adequate amounts of black rich soil, with sufficient moisture, and air holes had been punched in the containers. The covers of the containers were securely fastened to prevent escape. In the process of moving equipment on the workbench the containers of worms were placed unintentionally with one container resting against each end, or pole, of the magnet, and the other was a distance from the magnet. As the day progressed additional laboratory work cancelled the fishing plans. The earthworms and containers near the magnet were left in their positions for the remainder of that day and night until the following morning. The next morning there was an unexpected occurrence. The worms had eaten through one side of the container that was resting against the S pole of the horseshoe magnet, while the other containers were in no way changed.

The remaining earthworms were placed in fresh containers, again in the same positions in front of the magnet's poles—leaving one container away from the magnet. It was anticipated that if the poles of the magnet had any significance to the worms in the one container next to the S pole eating their way to escape, the result would repeat itself in another day. This did not occur. The experiment was forgotten for other laboratory work until three days later. At that time it was discovered that the earthworms next to the S pole had again eaten their way out of their cardboard container. They were lying on the workbench, had lost their moisture, and were dead. The phylum annelida, earthworm species, opened the door for further investigation and research.

Prior to this time there had been research in the laboratory to determine if electromagnetic energies had any effects on small animals, with no promising results until the earthworm incident.

In reproducing this experiment today it is noted that containers for earthworms are of heavy wax construction or other sturdier material than available in 1936. Experiments with the usual wax container on today's market take seven to ten days for the earthworms to chew their way to escape from the S pole energy field of a large horseshoe magnet.

By further experiments of trial and error it was discovered that the size and strength of the magnet and the temperature of the room surroundings contributed to the length of time for the escape of the earthworms.

For subsequent experiments the project of the magnetic worms received containers marked N for north pole, S for south pole and C for control. The control container was always placed out of reach of the magnetic fields of a magnet. Fresh soil and a few drops of water for moisture were added in each experiment and also a few dried leaves for food in each container.

Using heavier cardboard containers to prevent the earthworms escaping, the following results were after a twelve-day period of exposure to the magnets.

In the S pole container the earthworms were still present and alive, though they had been very busy chewing on the inside of their container. They were approximately one-third larger, longer in length and larger in diameter and were extremely active. Evidence of young worms in the soil showed a number of babies had been born.

The N pole container produced different results as many of the earthworms had died and those still alive were thin and showed little activity.

The control container showed no difference one way or the other.

The room temperature of this experiment was approximately 65 to 70 degrees Fahrenheit, as the earlier first incident occurred in a room temperature of 80 to 85 degrees Fahrenheit.

The magnetic worm project was continued and many experiments conducted for accuracy in results, always with the undeniable conclusion that the separate poles of a magnet had a different yet deciding effect on the subject. Investigations were broadened into other living systems, and laboratory records are available although too extensive for this accounting on the different effects on other living systems. The discoveries enlarged since the year 1936 are in many instances classified, and other developments are in further research before their publication. The classified material and continuing research show much optimism for great advancements in all fields of scientific investigation into the better understanding of the behavior and nature of animals and man and the world in which they live.

Our principal laboratory in Florida and our associated laboratories with other scientists under our direction have continued to release new discoveries in magnetism for the scientific community. Many of our discoveries considered nonsense years ago are now in use throughout science, yet there still remains in the scientific community a lack of appreciation as well as a lack of understanding of the true nature of magnetism and its application in our present world, both in health and industry.

As the magnetic worm project continued there was no doubt as to the totally different results obtained by placing the worms in the different pole fields of a magnet. Each test became more technical and time was lengthened or shortened. The strength of the first magnet used was 3000 gauss. The close relationship of each pole in a horseshoe magnet was not conducive to more accurate results in distinguishing the different effects by the S or N pole energies. Regardless of this, the experiments were significant.

In later research long, straight bar or cylinder magnets were used and this allowed a far greater separation of the poles and far more effective results. This type of magnet is now used in all tests and experiments in our laboratory research. Chapter One of this book describes the energy separation of the poles in various shaped magnets and presents certain graphic differences.

During the past few years scientific publications have announced a number of findings related to magnetic fields made by U.S. scientists. One such article in part disclosed that Drs. A. A. Boe and D. K. Salunkhe, two horticulturists from Utah State University, placed green tomatoes inside a magnetic field and discovered that they ripened four to six times faster when exposed to the S pole of a magnet or the open end of a horseshoe magnet. No mention was given to the effects or work done when applying the tomatoes to the N pole of the magnet since present accepted concepts of magnetic fields still rely on homogeneous, the same, which is incorrect.

THE CHEMICAL AND BIOLOGICAL ANALYSIS OF THE TREATED WORMS AND THE RESULTS

The study of the biological effects to the earthworms of the two pole magnetic fields was supported by the fact that the protein that makes up about 90 percent of the earthworm's system contains many types of amino acids. This indicated a sharp rise in the amounts available and also a reaction that caused unavailable proteins to become more available to the earthworm's system. The total results of the above indicated that acceptance of almost a total protein exchange was taking place in the subject's system.

These protein amino acids were an indicator of a form of life exchange encouragement to the worm's system, body, and physical development.

The S pole's magnetic energies had affected the sharp rise in protein amino acid development and active transfer to physical strength and developments. The N pole treated worms presented the findings that, unlike the S pole worms, the N pole worms were acted upon to reduce food intake, lessening the protein amino acid exchange, closing digestion of the lowered food intake, and this affecting a lower exchange of amino acids to physical strength and/or development.

The control worms, untreated, presented the same normal protein amounts much higher than the N pole's reduction effect and much lower than the S pole's treated worms. This index curve continued to present itself with the power energy curve of the limits of high to low gauss strength of the applied magnets and their pole energies. Where the low effects were sought, slow, longer time of treatment was necessary. Gauss of 100 to 300 were found to be the lowest effective energies or strengths preferable. This would enable a reproduction of the experiments with the same results time after time supporting this as a scientific discovery.

The highest level of energy found to be effective was 3,500 to 4,500 gauss. Above this the effects changed and even slowed in the effects that occurred. These experiments and further research showed a curve of effective strength that will then result in the highest degree of changes to any and all living systems placed under or in these separate fields.

It should be noted also that the waste matter discharged from the bodies of the worms contained a sharp rise in oils and fats and certain proteins as a result of the S pole or positive energy being applied in the prescribed manner to the subject.

When the magnetic energy was lower than 100 gauss, at levels lower than the earth's present one-half gauss of magnetic fields, very harmful effects were noted on the subject.

THE MAGNETIC EXPOSURE OF SEEDS

Hundreds of experiments were conducted at our Florida laboratory located in Green Cove Springs on the magnetic exposure of seeds. The results here proved to be another outstanding series of biological discoveries. The seeds treated before planting responded as did the earthworms—larger plants as a result of the seeds' exposure to the S pole and smaller plants as a result of exposure to the N pole of a magnet. The control, untreated, seeds acted as a guide and reference as to the opposite effects that were the results of these experimental magnetic treated seeds' growth and development.

The biological and analytical testing of the seeds at various stages of germination and development plus plant growth and development stages allowed even a greater understanding of such development results, such as the use of oxygen results and other results.

Exposing the seeds to the magnetic fields of the S pole and the N pole from eight to ten hours, to 80 to 100 to 280 hours, gave a great range of effects. Shorter periods of exposure from one to four hours did not effect the changes as much as the longer time periods. Quick exposure from several seconds to several minutes to one hour prompted certain improvements when the seeds were exposed to the S pole. Here we found the same reduction in strength and energy when the seeds were exposed to the N pole. The overall curve of graphed effects does not differ too greatly when the same time and strength are used to expose the seeds to either pole's energies. The effects in each and every case follow the same resultant pattern.

In these experiments the seeds were placed in small envelopes, the exact size of the pole's diameter, with the seeds lying flat in the envelope. The envelope was taped on the end of that pole of the magnet, marked and so identified. The control envelopes were kept in another room, far removed from any possible effects of the magnet's energies.

There was found to be marked differences when one group of seeds was treated for seven hours and another of the same kind of seeds for eight hours. Length of exposure is of the utmost importance in treating each type and kind of seeds. Radish seeds were selected for the first group of experiments, round, red types, as radishes germinate and produce a product quicker than other types of seeds that produce plant and vegetable products.

At various stages of germination, growth and development, laboratory conditions as to atmospheric and other environmental controls were carefully watched to insure an accurate result that could be reproducible subject to certain planned and controlled experiments.

INCREASE OF IMPORTANT PROTEIN, SUGARS, OILS, FOUND AFTER PLANT SEED DEVELOPMENT WHEN SEEDS ARE EXPOSED TO THE SOUTH POLE MAGNETIC ENERGIES

Laboratory analysis revealed the following. When exposed to the S pole energies the seed plant development to the end product, vegetable, fruit, root plants such as sugar beets, and all others planted, checked, replanted and harvested many times indicated that the plants produced remarkable results from the positive energies exposure of the seeds. The S pole energies tended to show rise in temperatures. Oxygen was liberated at over normal amounts. Intake of carbon dioxide was increased. Acceptance of organic matter, fertilizers, was increased and root products were greater. The length and size of roots were longer, having also a wide range in growth under the earth, and cycles where growth was speeded then slowed, unlike other untreated plants used as controls of the same types and kinds.

Sugar beets yielded more sugars. Peanuts presented outstanding increases in oils. Protein in the amino acids indicated increases as to the plant type and kind over normal amounts shown in hundreds of seed treatments, plantings and harvestings.

The opposite results occurred when the N pole energies were used to treat the seeds. This presented stunted growth patterns, products less than normal in all activities in opposition to the effects of the S pole energies.

Therefore, we have two types of energy—one that arrests life, growth and/or development, and one that increases life, growth and development.

The S pole or positive energies effects on the seeds show there are advanced and quite noticeable cycles to the growth and development of the plants. On planting there is a rapid germination period, then a period of rest where no development is indicated. On checking the root development there is a marked rise in root production. The top or surface development of the plants slows, then speeds up in very remarkable advance stages, not at all like seeds not treated or during their alternate periods of cycles in their development. Here we find another change over the norm of plant growth and development.

Again we find a similar effect from the two pole effects as seeds radiated within the two pole's energies. The product yield depends on the time and environmental surroundings of the plants during growth and development. The outstanding fact in research of tomatoes indicates that we could produce a tomato with less acid which as a result could be eaten by the many people who cannot eat usual tomatoes due to their high acid content. This lower acid effect is not due to the lowering of the other vital chemical contents of the tomatoes but is a result of a genetic change of the biochemical development of the tomatoes themselves. The experiments mentioned above were again obtained with the use of the S pole positive electronic energies only.

The S pole magnetic energies when used to radiate the tomato seeds produced tomatoes with even higher acid content than the untreated or control tomatoes. The use of the N pole to the tomato seeds prior to planting results in a less acid tomato. The resultant effects of the seeds in a number of cases reverse the effects one may expect as a result after radiation of either energies as to plant content, the biochemical constants.

Chapter Five

THE TWO POLE EFFECTS ON
SMALL ANIMALS, SNAKES AND BIRDS

In the introduction of this chapter we would like to present an outline of one very important experiment where a magnet became a mother to a group of baby chicks.

We believe this to be an outstanding discovery that deals with the inborn intelligence of small animals, birds, and in this case newborn chicks still wet from the egg. The highly inborn sensitivities and the psychological reactions proved to be very unexpected and accidental, yet important and very rewarding series of research findings.

Moving from earthworms to seeds, laboratory research was focused upon altering any aspect in the development of small animals, as this would be further proof of what the two pole discovery had to offer mankind. Eggs of the normal white leghorn chicken were chosen. Taking two dozen fertile eggs they were treated in groups of eight to each group. One group was treated with the N pole energies, a second group with the S pole energies, and the untreated group was kept away from the magnetic fields. All were kept under laboratory environmental controls. The untreated group were the controls.

For the S pole energies each egg in a group of eight was placed in front of the S pole of a 2500 gauss magnet, using a separate magnet for each egg and placing it in front against the S pole. The eggs were turned every three hours. The same procedure with the N pole energies was used with the second group of eight eggs, and the third group of eight eggs was placed well away from either pole's energies. Magnets used were straight cylinder magnets. Temperature during treatment was steady at 80 degrees F. Three small electronically controlled incubators were used, one for each group, to facilitate treatment at the same time under exactly the same conditions except for exposure or nonexposure to the pole's energies.

The incubation period was two or three days sooner than the normal time with the S pole treated eggs. The N pole treated eggs were slower to hatch, from one to two days. The greater importance of this experiment was yet to come; on the removal of the chicks exposed to the magnetic energies and placing them in suitable cages, each group in separate cages under the same environmental conditions, a horseshoe magnet of about 5 x 6 inches with a pole distance of 2½ inches was placed in each cage. Also, a dummy magnet made of wood of the exact same size and painted wtih the same paint as the real magnet was placed in the cages. The wet chicks just leaving the eggs were immediately transferred to the cages containing the magnets. Each cage had water, baby feed, and floors covered with soft white paper. The chicks in the cage marked S pole treated, as soon as they were half dry from leaving the eggs, took turns and one at a time entered between the poles of the real magnet only. Each chick would remain between the poles of the real magnet about two minutes, then leave and retire as far as possible from the real magnet. Then another chick would enter and reenact the same process. This continued until each of the chicks had entered and lain down within the two poles of the real magnet, rested for two minutes, then left the magnet. Not one went near the wooden dummy magnet. This was a lesson and a discovery as to the inborn instinctive intelligence of the baby chicks. This experiment was repeated many times using eggs of other breeds of chickens. The reenactment was exactly the same in each case. Their inborn intelligence acted to attract them to the magnet as a chick would seek out its natural mother for heat and comfort. This psychological intelligence did not come from experience or prior training. It was quite clear that the chick's natural instinctive reactions sought out and directed it to a source of strength and comfort. However, giving strict attention to the time each chick remained in the fields of the magnet allows us to see that the chick was aware of the intensity, power, energy of the magnet, and its inborn sense reacted to time the exposure to the magnet, then leave that energy source and travel to the farthest point possible within the confines of the cage. This timed the amount of energy that the chick's inborn system told it was enough. These series of experiments were termed The Magnetic Mother. It was clear the chicks identified the real magnet's energies as a comforting, strengthening source much the way they would seek out and stay within the protection, warmth, and energy provided by their natural mother. The N pole treated chicks stayed in the field slightly longer, for periods up to three minutes as they were reflecting the arresting, limiting, reactions of being treated by the N pole fields prior to incubation. The control chicks waited until they were dry of their shell's wetness before entering the magnetic energies. Time was longer than the S pole treated chicks. The chicks from the control group seemed to need more energy. They remained between the poles of the magnet from two and one-half to three and one-half minutes before leaving the magnet for a remote area of the cage.

The experiments with the chicks were important laboratory findings in the use of small newborn animals to detect changes in the normal attitudes, behavior, intelligence, psychological behavior, mental activities, and developments. At this time, and as a result of many subsequent experiments, laboratory findings indicate it is now possible to program degrees of intelligence in not only animals but also man by the proper controlled use of regulated magnetic fields of energy. Some of our laboratory findings in this regard will be discussed later in this book.

THE GROWTH AND DEVELOPMENT OF THE TREATED CHICKS TO CHICKENS AND ROOSTERS

Watching and recording with great care the development of the chicks into hens and roosters brought many new and important developments and discoveries. The development of the S pole treated chicks—hens and roosters—presented these facts. They grew faster and stronger than the N pole chicks. They ate more and near maturity took on a trend toward being cannibalistic in nature. Their intelligence was lower in all respects than the other two groups of chickens and roosters.

The N pole treated chickens and roosters were light eaters. They developed slower than the control chicks. They were sensitive to all surrounding noises, heat, cold, wind, sun, weather. This was opposed to the boldness, dull thinking and reactions, and overly strong S pole chickens and roosters. The S pole hens and roosters were indifferent to any surroundings when their behavior was compared to those of the N pole. The control hens and roosters were in every respect normal to the accepted behavior of hens and roosters. The great differences the birds presented was in fact outstanding. As in the experiments with the earthworms and with the seed experiments, similar findings yet different developments took place. The chicks were studied from birth to maturity to death. The S pole roosters during the last stages of maturity attacked and ate the flesh of the hens and their own kind. It was necessary to remove them and place each in a separate cage. The sizes were much larger than the control chickens. The N pole species were thin, nervous, very sensitive, very clean and ate sparingly. This group was completely different from the control group, which were active and scratched for their own food, and drank less water than either of the treated types. Their growth was larger in all respects than the N pole group and far less in growth development than the S pole group. The S pole treated group (eggs to mature hens and roosters) were the leaders in the cannibalistic attitudes. The birds accidentaly left the confines of their cages a number of times when the helpers failed to properly latch the pens. They were found running dogs, cats, and in one case attacked a cow grazing in a nearby pasture. The attacks were all of the same nature—mounting or flying on the animal's back and laying open the back. In their own pens when this was discovered for the first time, it was believed that an animal had somehow got into the pens and killed two of the large hens. Upon careful examination and watching the attack was repeated. The S pole roosters mounted the backs of the other birds and then proceeded to peck, scratch and dig into the center of the back of the other birds, exposing the internal organs, and death then was the result of bleeding and internal organ damage.

During the last stages of development the N pole treated birds lowered their water intake and increased their food intake. This made no difference to the weight, showing here the control effect of development of the birds so treated at conception and prior to conception. The effects then followed through the stages of development from the embryo. These experiments were reproduced many times. It was very clear that certain genetic changes affecting the growth, development, attitudes, physical development, mental attitudes, and psychological attitudes had been altered or changed by the magnetic separate pole radiation of the eggs.

Now we move to a new series of tests encompassing the use of mice and rats with generally the same effects, although these experiments presented new facts in the development of abnormal sex encouragement.

THE EFFECTS ON AND TO MICE AND RATS

In the treatment of mice and rats we built suitable cages to allow treatment of the males prior to intercourse with the females. This was necessary so the sperm would carry the pole effects and the active transfer of the sperm could be timed, checked and recorded.

Three groups of white lab mice were carefully selected. One group acted as controls and the other two groups were marked S and N treated mice respectively.

The male mice were treated in a single cage, one to each cage, with the S pole energies. Here 2500 gauss was used for eight hours. The cage was designed to keep the subject in the S pole energy of the large 2½ x 6 inch bar or cylinder magnet. At the same time in separate cages, well away from the S pole treatment cages, the male mice were treated with the same type and strength magnet except with the N pole of the magnet described.

We then placed the males and females together and normal intercourse took place. In a few weeks we again saw a shortening of the time the babies were developed and delivered by the S pole treated mice. Again, new and important findings were made. The births were easier and the babies were larger than the N pole treated mice. The controls were the same as normal delivery. The N pole treated mice babies were more difficult to deliver than the controls. The babies, as in the S pole treated mice, were larger and in some cases took longer to deliver.

The S pole mothers were stronger, and less effort for delivery was noted. This was opposite to the N pole mothers, which seemed to be lower in strength, and the babies were smaller compared to the controls.

Before the birth of the mice each cage was equipped with

separate huts and with two openings. This allowed the mothers to protect their young and keep them warm and away from any source of danger. While this danger condition was not anticipated, the care and safety factor had to be noted and was made available to each mother.

On birth, the S pole babies developed faster than the controls. The N pole babies took longer to develop, were weak, thin, and did not feed as much as the controls. The S pole babies were fed continually. They were stronger in every respect than the controls.

The same experiments were conducted with white lab strain rats, which are similar to the white rabbit strains and come close to the blood system of man.

The rats followed the exact behavior in development stages as the mice. It was noted that one important result of the mice and rat tests showed that the control mice kept only a fairly clean cage, nest, and hut. The S pole treated mice and rats kept their cages, nests and huts in a very dirty state; they did not seem to be concerned about sleeping in their own mire and filth.

The N pole mice and rats were very neat housekeepers and often took a great deal of time washing and keeping themselves and their cages clean, including their nests and huts. On the other hand, the S pole mice and rats were always stained, dirty and careless. The controls were not extreme one way or the other, their cages, huts, nests and cleanliness ordinary for their species. The apparently high sensitive behavior to lights, sounds, motion, movement in the laboratory, by the N pole mice and rats as opposed to the boldness, strong, nonfearful behavior of the S pole mice and rats showed a remarkable difference in the psychological behavior pattern of the rodents. These differences coincided with our findings relative to the particular pole energies.

The discovery of these revealing changes in the mice and rats was duplicated in the work that followed with white Australian rabbits, which have a blood type similar to man. The effects were so far-reaching in this work as to point directly to genetic changes. This was shown by the sensitivities, physical development, nervous reactions, and the trend toward cannibalistic behavior. There were also effects on the sexual behavior of the rodents.

SEX LIFE AND AGING

The sex life of the mice and rats of the control groups was considered to be normal and we used their behavior as norms.

The sex life of the N pole rodents was limited and less active than the controls. It was noted that experiments with the mice, rats and rabbits all resulted in the same percentage of exactness in resulting behavior. The S pole rodents, encompassing all of the above-mentioned types, reacted to a far greater sex life with frequent activity. In some instances the males killed the females by their sex activity of actual viciousness.

The exposure of the rodents to the S pole energies acted to inspire strength and vigor and when applied to the sex organs encouraged their overdevelopment. This was also later discovered in cats and dogs. The amount of sperm produced and the larger percentages of resultant fertility were responsible in part for changing the rodents and animals in their inborn habits, personalities, behavior, and encouragement of sexual activity and reproduction.

In treating animals after maturity that had not previously been subjected to the pole's energies, the result was increased strength and sex activity. These experiments were by exposure to the magnet for one hour a day for four days, the curve of effects varying with the size and type of animal or rodent. The exposure was to the male testicles and to the female reproductive organs—S pole 2500 gauss strength.

The result, if left unchecked, of the condition of oversex stimulation was to shorten the life span of the rodent or animal. The heart was affected, shortening the life span, and death resulted.

In our research of rodents and animals to arrest the oversex activities, it was found much could be done if the N pole energies were directed to the male, exposing the testicles and the ureter with N pole energies one hour a day for three days. This resulted in a noticeable downgrading of the number of sex acts that were performed in a definite period of time.

Of equal interest during these experiments was the measurement of the amount of sperm produced by the male animal. In treating the sex organs with N pole energies less than normal amounts were produced. We again used a similar animal as control—one that produced the same amount in close percentage to the animal selected for the measurement experiment. Here we note that the sperm is for the better part protein; therefore, should the exposure of the S pole produce more measurable sperm after a series of exposures to the animal we can then see that this acted to encourage the production glands to effect a higher production of the protein sperm fluids.

Quite the reverse was found in the treatment of animals with the N pole energies. We found a sharp lowering of the production of sperm and a lowering of the amount of protein.

We can see the possibility for the same reactions taking place with man, since man and these selected animals have the same or similar organs.

THE INCREASE OR REDUCTION OF THE NORMAL LIFE SPAN OF ANIMALS

The life span of rodents and animals can be extended up to 50 percent. Mice and rats proved this possibility. In larger animals it has been more difficult to note this due to their normal life span reaching 18 to 25 years, as in the case of cats and dogs.

However, the larger rats, as did the smaller mice, allow a good reference in themselves and in comparison to the normal life spans in other animals when they were treated with magnetic energies before active transplant of the sperm, after the transfer, and/or during conception of the embryo. On applying the S pole fields prior to the first stages of development, on birth the animal carries the changes effected by the S pole exposure. If after birth of the rodent the applications of the S pole energies are made there is a lowering of the life span due to oversex results.

It was concluded that if the animal's genetic mode is altered to one of a higher strength, the life span would be extended by the fact we have altered the genes and the resultant strength of the rodent. The main and most difficult matter to cope with is that due to the oversex resultant condition to extend life one must isolate the male animal or rodent for lengthy periods from the female or place restraints on the sex activities of the male. If not, then depletion of strength acts on the heart and organs and reduces the life span one may expect by many experimental results.

The N pole rodents and animals also show an extension of life but by different approaches. The use of the N pole energies to extend life is quite different. During these N pole experiments it was discovered that the extension of the life span was the slowing down as opposed to the strengthening of the rodent's or animal's system. This presented a slowing of maturity, thus resulting in a longer life. This should open many new avenues of research as in each experiment the perception and intelligence of the rodents and animals were upgraded as a direct and positive result. The N pole exposures resulted in a weaker, smaller rodent or animal of slow development by extending its normal life span and upgrading all sensitivities, including intelligence, reflexes and environmental reaction, inferring the brain's ability to be more sensitive in recalling information and environment.

The N pole animal or rodent was then more intelligent than the dull-witted, overly strong, slow to learn, animal or rodent we have found is the result of the S pole exposure. The strength-giving results of the S pole energies and the resultant changed animal while being overly strong was in no way slow to move and respond to activity, yet there was a failure to have the quickness of mind that the controls presented or the increased mental activities of the N pole animals or rodents.

These were the results in each of over 300 experiments conducted within an eight-year period in our laboratories. From these experiments and their results, a reasonable possibility exists to program certain and very advanced degrees of intelligence to rodents and animals and, therefore, within the possibility to consider the same for man.

There is a similarity with these experiments and the proverb "the wiser, the weaker." Concerning the changes and effects in the rodents and animals this was the result of the application of the N pole energies.

The extension of life systems to live beyond their normal life span has received much discussion in scientific reviews. Men and women advance in their development to become outstanding authorities on important subjects and then die. If they had lived longer more information from their efforts would be available to aid mankind.

The laboratory work with rodents and animals indicates that the life span of man can be extended. Yet, who will decide who is to receive and who is not to receive such treatment, if perfected? What effect on society by the have and the havenots would there be? The restrictions upon these men and women that were programmed for extended life could be impractical.

Laboratory results indicate that the proper magnetic fields applied could aid man as they did the rodents and animals. The results also indicate the opposite effect with improper or opposing magnetic fields. Many diseases suffered by man, simulated in laboratory experiments with rodents and animals, were contained by applying magnetic forces. The possibility of strengthening the heart and mind of man and arresting illness and disease exists with the proper use of magnetic energies. In the many years of working with magnetic energies on animals, our laboratory has arrested illnesses and diseases in more than several hundred subjects. Many of these ailments are common to man, and a number have not responded to modern medical treatment.

In scientific research we try at all times to avoid duplication of work. Our studies have shown for some time that vast improvement in the animal's circulation of blood can be obtained by the proper application of magnetic energies. The scientists in Russia have also found this reaction to the application of a magnet's energies. However, the scientists in the Soviet Union are working with both poles of the magnet at the same time, and they consider that the two pole energies are homogeneous. We have discovered that the use of each pole when properly applied has presented us with a go or no-go, in computer language, method and/or system to work with, using the two different electron spin potentials for better results. The approach of using the two poles at the same time should not be discounted as they are a very valuable tool in that form of energy presentation, but this approach is not as accurate as the two separate pole system which can be computed and programmed for desired effects.

As a result of our discovery we can attack the cause of poor blood circulation and relieve the condition, if we know what caused the condition to exist. In any case of abnormalities to the system, knowing what caused it to exist and why it is happening is one of the utmost importance. Having an analysis of the condition, the energy that will effect an arrest, control or relief can be properly applied and a certain degree of results may be expected and generally occur as expected.

Much can be done to relieve many heart disorders. This and the possibility of relieving certain types of kidney disease can now be accomplished in blood-circulating animals. Because of the similarity of the blood and organs of animals to those of man, we can clearly see the possibility of our research in relieving man of these certain ailments and diseases.

The possibility of aid and relief of many liver complaints has also been indicated as showing positive results. These findings show in part the aid to the extension of the life span of man, as death is caused to a large degree by the above-mentioned illnesses and diseases.

While heart failure may be considered as the main reason of death in man, this is due to the life system becoming overworked with worry, nerve reactions, loss of strength, and the aging process of man. When the total system slows down, life slowly comes to an end.

Laboratory findings indicate we can, to a certain degree, arrest this weakening condition by the use of the energies we now have to work with—those that have shown that animals can live longer. Even mature rodents can be acted on to extend their life when these energies are used to reinforce to a degree their strength and protein exchange of the foods eaten. All this acts to assist the heart's actions and improve circulation, thereby removing a good degree of the cause. This and other measures may be taken to aid the living system by retarding the aging process.

The physical appearance of a rodent exposed to the S pole or the N pole energies while very young or during or before conception presents the picture of extended prime life even at the end of the period that the rodent's life span may be considered at an end. The appearances are outstanding in every respect. The fur is that of a middle-aged to younger rodent or animal. Aging has been retarded. There is no doubt in our minds that this is the result of exposures to the magnetic fields. Too many reproductions of these research experiments have been duplicated to conclude otherwise.

In man, as in rodents and animals, as age progresses many systems indicate loss of certain mental activities, sensitivities and interests. In our laboratory experiments we have seen these actions delayed, arrested, and much activity restored—all important findings.

Would extending man's life span overburden the earth too much to allow life as we now know it? The reason health matters are growing worse is that too many people are in crowded conditions. The cities are overpopulated. The world has millions and millions of acres of unexplored and undeveloped land and has the potential to properly feed all the world. Still, thousands die each year in poverty and starvation. Should we consider means of extending the life span of man it would require an international planning board to open new lands and new housing developments. It would demand increased national development of many nations of the world that today cannot provide sufficient food for their people or proper housing.

In the U.S. alone there are millions of acres of undeveloped land. We see the possibilities of great new government and industrial areas of new and promising developments. This also applies to many nations that are not in the development stages of America and other well-developed countries. The world is now undergoing drastic changes, and there will have to be better planning, more properly educated men and women in government positions to effect these developments. Wars as we have known them must stop, and sensible approaches must be applied to end world conflicts—man against man, nation against nation.

Today, and this includes all nations, countries, states, and also the U.S., it is doubtful if the men and women in political leadership are suited to handle the vast new concepts and developments the world needs for proper, sensible government and sensible development of their resources. Extending the life of man is a possibility if properly planned and committed to action.

As scientific advancements are made, the people of the world also change in their daily lives, eating habits, housing, activities and comforts, yet we see little improvement in government procedures directed in keeping up with the changes all nations face today, other than reacting to crisis after crisis. The present trend to social controls has never worked as people must have the right to think and act for themselves, and this is fast coming to an end in all countries of the world. The same is necessary in research and development.

The world today is uneducated in keeping current with scientific developments that are made to aid the people of the world. The political leaders are not informed on scientific breakthroughs or, if they are, for unexplainable reasons, they do not properly follow and aid these breakthroughs for the betterment of mankind. Passing more laws and placing greater regulations on their people by all nations is not going to be the answer or in any way provide us with the answers we need now more than at any other time in history. We need a fresh, new approach to science and its uses for mankind.

Chapter Six

MAN THE ELECTROMAGNETIC ANIMAL

We will investigate in this chapter the facts surrounding man and his biological electrical energies. Man is composed, as are animals, rodents, seeds, plants, and all biological systems, of the basic atom which in part is a small magnet. As atoms collect to produce molecules and molecules form to make materials, substances, matter, all are allied and are, in fact, the fundamental basis for man's existence. They make up man's body, system, and are his very life itself.

The electrical system of man and all biological systems is a complex electrical carrier, some still unknown to man. As research continues we are becoming aware how these electrical currents govern man's life and the lives of plants, animals, rodents, and other living systems.

Of all the complexes of man's electrical biomagnetic system the brain and mind stand out as the master computer. The brain and mind are an electrochemical master control and exchange system acting to govern a good part, although not totally, of the electromagnetic complex found to exist in living systems.

The use and application of a magnet's energies, magneto magnetic, biomagnetic energies, to the living system can and will effect certain changes. When we know how these affect man they can then be directed and programmed to assist in the correction of many complaints existing in the electrochemical system of man's organs, blood, nerves, heart and all parts and divisions of man's biological living system. This will be a tremendous step in the control of illnesses or human restrictions toward living a more normal life when afflicted by illness, disfigurement, loss of limbs, or other limiting factors.

It would require a number of books to elaborate in detail all of our experiments and findings in the bioelectrical activities within the body of man. At a later time we may present another book with more of our research findings on the magnetic effects to the living system.

In all scientific investigations when a discovery is made it is of the utmost importance to understand why it happened, not simply apply it for the results it can afford. Making a true discovery in science is good, but to properly understand what your discovery does should be of the utmost importance. We mentioned earlier that the Japanese are making magnetic bracelets and prescribing them for all sorts of human ailments. While the bands have made what appears to be certain improvements in a number of cases when applied to the human body, the Japanese manufacturers and scientists lack the knowledge of why this has taken place. This is evident based on papers they have offered on medical findings. These medical findings lack the proper scientific investigations into the technical reasons why these improvements have been experienced.

Actually, you can take a child's magnet and place it at points on the human body and obtain reactions, certain amounts of effects to certain disorders. However, unless you understand why this takes place, then nothing has been done to advance the position or understanding the laws behind this effect. We hope to cover this in part in this book.

While our research has included many fields of investigation over many years, we have primarily restricted our major work to animals and biological systems. When we apply our findings to the work being done by many other dedicated scientists in this and related fields, we see there is a direct relationship between our work and the work being conducted by other scientists. The research results we present in part should provide a valuable tool as to experiences, experiments, and results obtained for more thorough understanding of new basic laws we have developed from the effects of use and application of magnetic fields and energies. Only now are we beginning to grasp a more complex understanding for the application of these fields and energies to obtain desired and planned results.

MAN'S ELECTROMAGNETIC ENVIRONMENT

Man is affected by incoming electromagnetic energies as those coming from outer space, the moon, the sun, and the other major planets. These energies combine in various ways to affect man and every living biological system on the earth. Those creatures that live beneath the earth, in the seas, lakes, rivers, are all affected by these unseen and normally undetectable energies that continually bombard the earth.

The moon affects the rise and fall of the tides, as does the sun; and when the moon and the sun's energies are combined, when they are in alignment, they act to exert their maximum pull on the earth, producing abnormally high tides. Not only are all biological forms of life, including man, affected by the geophysical properties, but segments of the earth also undergo many forms of pull, strain and stress.

Man, like all living biological systems, experiences changes in these pulls, strains and stresses that are a result of these bodies from outer space and their surrounding electromagnetic fields and gravitational forces.

Man is an electromagnetic animal and is subject to those forces that affect all forms of life existing on earth. Man's electromagnetic system is contained within his biophysical makeup and affects the total behavior of not only the body but also in many cases the changes in mental activities and the electrical biochemical operation of his system.

We are aware of the numerous defensive shields surrounding the earth's atmosphere which protect man from the deadly rays of radioactivity from the sun's radiation. These shields are electromagnetic and have an energy that adds to man's environmental electromagnetic environment, adding also to the effects on his biological system.

In the past few years, gravity, as known on earth, has been discovered to travel in energy waves. Gravity is a form of electromagnetic and physical magnetic results. Without gravity our blood would not circulate the same as it now does. Therefore, each of the forces exerted on mankind on earth plays an important part in man's electromagnetic biological health, atmosphere and environment. When we add all of these external electromagnetic energies to man's internal electromagnetic energies, since man is himself an electromagnetic animal, we find man and his surroundings are subject to tremendous magnetic pressures, strains and stresses. If we can understand, even in a small way, how to harness some of these natural forces and apply them to aid mankind we have at least started to advance our understanding of these natural forces that affect the entire biophysical and biological atmosphere of man.

Nature presents some remarkable evidence of the effects the moon's and sun's energies have on living systems. Take for example the oyster. The following experiment has been conducted by a number of researchers who are exploring biomagnetic fields and their energies. Taking a number of oysters and placing them in tanks far inland and underground where they could not possibly sense outside happenings, the oyster still opened and closed their shells in perfect rhythm with the rise and fall of the ocean's tides. This occurred although the oysters were moved miles from the ocean. The energy force from the moon, the magnetic and electromagnetic energy of the moon's gravitational force fields acted to provide the oysters with the information, stress and strain, that caused them to react as though they were in or close to the ocean.

Consider that a magnet's energies cause a more rapid germination of seeds when exposed to the fields of a magnet and when planted present these facts repeatedly. Consider also that the seeds are organic in nature and have enzyme systems. Life, even in a suspended form, before the forming of the embryo, is influenced by electromagnetic environment. Any electromagnetic energy directed to a life system affects the protein, amino acids, and enzymes. Further, we see genetic changes as a direct result of magnetic exposures to the living system that are laboratory reproducible. An improper application of magnetic force to that system will produce mutant, defective genetic results, while the proper application will produce genetic changes for a better end result.

Research is now underway in several leading hospitals to show that man's brain is subject to severe changes when exposed to external minor voltages and currents. It has been found that the front of the skull is negative in electrical potential and the back of the skull is positive in nature. When small electrodes are attached to the front and back of the head and small voltages and currents are applied—positive to the back, negative to the front— there is a feeling of well-being. However, should we reverse the voltage and currents quickly, unconsciousness will result immediately.

Electrosleep is the result of applying different frequencies of alternating energy to the brain by small electrodes. This produces sleep until the currents of frequency application are removed. Electrosleep can be used to assist man or animal to sleep while undergoing minor operations. It also has its use in psychological treatment and other research applications. The Russians have pursued this type of research for years, and now the United States and other nations are finally researching its uses and applications.

The above types of energies and other forms of applied voltages, currents, alternating frequencies, magnetic, and electromagnetic energies are now being researched in laboratories throughout the world. Much has been discovered about the effects these applied energies produce on plant, animal, and man.

Magnetic fields that are lower than the earth's one-half a gauss are found in space after leaving the earth's magnetic fields. This energy assisted mice taken into space to grow faster, but the second generation of mice were found to have low vitality and strength and died young. Their organs were found to be affected when they were contained in a weaker magnetic field during their youth. Tests indicated clearly that their organs were poorly developed, their livers and kidneys malformed, and many developed malignant tumors in various parts of their tissues.

Here we see a clue to malignant developments that result in many types and kinds of cancers and tumors. In the study of cancers and tumors there is more than sufficient evidence to indicate that when a cell membrane that retains the cell's shape and form weakens it can be the first cancer cell. We will discuss this in more detail later.

A number of years ago scientists in the U.S.S.R. started to investigate the effects of sunspots and magnetic storms on human behavior, studying patients in hospitals suffering from blood and heart complaints. Their discoveries acted to create interest in the external electromagnetic effects on man. Interest is now beginning in the U.S. and other countries that realize there is a direct relation between sunspots and magnetic storms on the health of man.

The Soviet scientists designed shielded rooms and changed medication to the heart cases, those suffering nerve conditions and blood ailments. This procedure was used to protect the patients against electromagnetic harmful effects found to occur in cases from the effects of sunspots and magnetic storms.

It has been discovered that pulsing waves or fields of electromagnetic energy in factories and plants can cause great harm to the living system and also to the mind. Large transformers, pulse coils, autotransformers, A.C. generators, all show dangers to the workers. The pulsating generators may be considered to be the most dangerous of all, producing effects on the workers which result in harmful complaints. In 1969 reports were made public by Dr. Karel H. Marha, Institute of Industrial Hygiene and Occupational Diseases, Praha, Srobarova, Czechoslovakia showing effects to the workers. The complaints reported included decrease in sexual potency, headaches, memory and hearing losses, and changes in menstrual cycles. Additional human disorders have been noted and identified since that time, as well as several years prior to that time. The dangers of microwaves, such as radar, microwave ovens, and other similar devices, can also cause great harm depending on the energy leaking, the frequency and power generated, as well as the time of exposure.

The proper harnessing of these mentioned forms of electromagnetic energies can result in outstanding discoveries to help mankind. Our laboratory findings substantiate this objective.

It is possible to design from a harmful pulse generator a device to dissolve unwanted tissue. Tumors may be dissolved as well as other growths within or without the body. Our laboratory has made a number of research instruments that have acted to dissolve unwanted mass in animal bodies. From these research findings have come information and knowledge of how pulsing electromagnetic energies can be new and important tools for medical work, surgical procedures, and of equal importance to avoid surgical procedures.

As mentioned earlier a number of firms in Japan are building alternating current equipped chairs, mattresses, pillows, and hand-held devices that are placed against the body to afford a host of reliefs and cures, according to the literature they send with these instruments. These do-it-yourself, home cure instruments supply a 60 pulse per second magnetic, alternating magnetic energy, and can affect the system in many harmful ways. They have obtained many good results, meaning certain amounts of relief to a number of disorders, yet they do not properly understand that their instruments can have harmful as well as good results. This is indicated in discussing with them their papers and reports they offer in evidence of the instrument's value to health.

Our laboratory findings give proper understandings concerning these devices built by the Japanese manufacturers. An adaptation of these devices in line with our discoveries would not only benefit but would offer new and greater approaches in proper medical testing and results for new scientific instruments that could be accepted by many nations.

The research work on instrument development in England, France, and Germany is evident by a number of groups attempting to design, build, and sell magnetic instruments to relieve human complaints. Again, these instruments are based on the pulsing frequency type generators, generating electromagnetic energies and applying them to certain parts of the body for relief of certain disorders. A number of these instruments have obtained positive results, yet these manufacturers fail to understand the nature of the energies they are working with or how to program for expected results. This book is directed to professional men and women and students who wish to further their research of applied magnetic forces, fields and energies in an intelligent and understandable manner. Such persons must be willing to leave behind them the outmoded, incorrect theories and concepts of magnetism. Our laboratory findings and discoveries of new laws and concepts of applied magnetic forces can now be used to advance new developments and still more discoveries in advancing the sciences. The future holds many great and new discoveries in all sciences from our new understanding of magnetic energies and their programmed effects.

EXTERNAL ELECTROMAGNETIC FORCES AND MAN

In our continued study of the electromagnetic forces that are directed on mankind from outer space, we find one of the greatest effective generators of harmful, as well as useful and healthful, energies are the rays of the sun. The sun's rays are electromagnetic in nature and effects.

The sun's rays, on direct exposure to the skin of animals and man, can cause skin cancer if the exposure is prolonged. Sunlight and direct exposure to ultraviolet rays can also cause certain types of skin cancers. Why the sun's direct rays or the direct rays of ultraviolet light cause certain types of skin cancer is not too well known or understood. We point out that ultraviolet rays or light are also electromagnetic energies. The frequency of the sun's light rays, white light, and that of ultraviolet rays differ. The white light that comes from the sun contains all of the frequencies within the light spectrum and many that go beyond this range, and other frequencies that start before the sun's white light frequencies start. There is a difference in the number of cycles per second in each wavelength and micron. The difference is the number or the frequency of the cycles per second of energy that goes to produce, make, and/or present these many different types of electromagnetic energy.

We believe others will agree with this finding at least in part and that it is the sun's reactions and also ultraviolet rays' reaction on the skin of man and/or animals that cause "oxidized cholesterol" of the skin. We know there are many biological effects of cholesterol oxidation. One result is cholesterol alphaoxide. This is known to be a cancer-causing chemical, so when exposed to certain natural forms of energies coming from outer space, as in the case of sunlight, we can see there is a direct possibility that when too much of this energy form is absorbed by the skin the result is skin cancer or deterioration of the skin.

Man is continually bombarded with visible and invisible electromagnetic energies that have a direct bearing on his life, mental attitudes, health and welfare. Man is, in fact, an electromagnetic animal in every respect, living in and surrounded by an electromagnetic environment over which he has little control.

Chapter Seven

MISTAKEN CONCEPTS OF MAGNETISM AND ITS BIOLOGICAL APPLICATION

We now present some of the theories of the magnetic fields that surround the earth and a magnet. We give a drawing of the earth, Diagram A, that shows a bar magnet superimposed on the earth, and in turn shows how the magnet and the earth's magnetism are similar in poles and flow direction of the magnetic energy. This theory is incorrect as we shall explain, yet it is still taught today in this incorrect form.

DIAGRAM A

DIAGRAM B

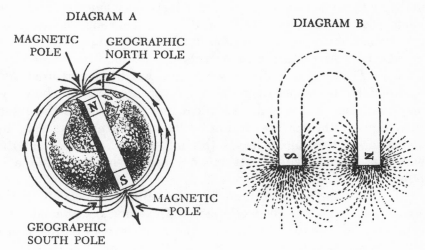

MAGNETIC POLE

GEOGRAPHIC NORTH POLE

MAGNETIC POLE

GEOGRAPHIC SOUTH POLE

This mistaken concept is used today in most biological research dealing with the science of magnetomagnetic effects to biological systems. Diagram B shows the mistaken concept of the magnetic field of a magnet by the paper and iron filling test. This consists of taking a piece of paper, placing a magnet under the paper, sprinkling some powdered iron on top of the paper, and the

resulting arrangement is supposed to show how magnetic energy
flows from pole to pole. Since each filing particle becomes a
magnet, as explained earlier in this book, this method and deduc-
tion are in error.

We now present the updated theory of magnetism surrounding
the earth and also a bar magnet. We refer to the measurements
made of the earth's magnetic fields as recorded by the research
conducted in space by the National Space Administration's research
of magnetic measurements. This compares with our findings that
the earth, like a bar magnet or any magnet, has a magnetic equator
and it is at that point where the spins of electrons change their
phase relationship and present us two fields of energies and two
different potentials of magnetic energy. This offers a totally different picture than is now used in present textbooks and is used
as law and theory in all related research. Our laboratory findings
also show where the magnet's energies should be applied in and
to biological systems to study the effects of magnetic forces on
living systems. For the study of the new concepts as to phase
spin change and relationships refer to Chapter Three.

The two drawings shown on page 57 are marked Item C and
Item D. Drawing C presents what now has been measured and
recorded by the space recordings and probes. We see how the
magnetic energy leaves the earth's S pole, spinning to the right,
then dips to the surface of the earth, and changes its spin by
180 degrees, then again leaves this mid-magnetic equator of the
earth and travels with a left spin to reenter the earth at the N
pole.

The drawing marked Item E is an outline projection of the
Van Allen radiation magnetosphere. This envelope-shaped field
contains radio active fields and many other atomic radiations.
Do not confuse this field with the existing magnetic fields we discuss. Item D shows the same magnetic equator that is present
and is shown in Item C. The bar magnet has this division of
energies as does the earth. All magnets have this magnetic equator
where the energies are divided and changed as to their magnetic
spin effect, which then presents us with two values of magnetic
energy—south, or positive, and north, or negative, energy. It is
the popular belief that magnetism flows only in one direction.
However, we have presented evidence to support that a magnet's

ITEM C   ITEM D

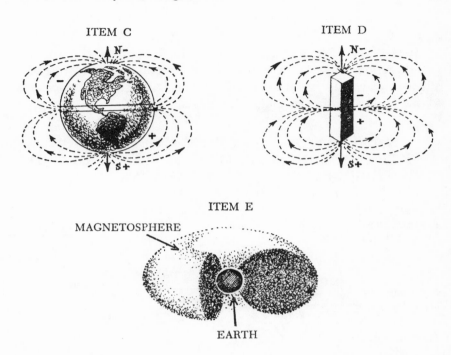

ITEM E

MAGNETOSPHERE

EARTH

energy has two flow directions. A magnet's magnetism flows in two directions, not one. This follows the concepts of electricity that voltage and current flow in opposite directions. Here we again see an almost identical behavior of magnetism and electricity. In fact, we cannot have one without the other.

As we continue this book we will show how the two energies, voltage and current, that we have in a single pole energy, either the N pole or the S pole, can be made to perform with different yet almost computerized exactness in obtaining the results desired in treating biological systems.

By application of the N pole's magnetic energies we can arrest certain bacteria while strengthening the normal body cells surrounding the invading and attacking bacteria. Yet both are living systems. Where E is the voltage effect and I is the current effect we then have E and I in a single pole energy that applies equally to the N pole, negative, and the S pole, positive, energies.

Refer to the detailed outlines and drawings used to support the fact that the flow of magnetic magneto energy from the poles of any magnet has a dual nature, flow, and type of energy, again E and I.

It is of extreme importance that this simple interaction of a two-directional flow taking place be understood. It has two general effects to each biological application, each having great importance to our understanding of magnetic effects.

A LABORATORY DEMONSTRATION TO SHOW THE TWO DIRECTIONS OF FLOW OF POLARIZED HYDROGEN BUBBLES IN A MAGNETIC FIELD

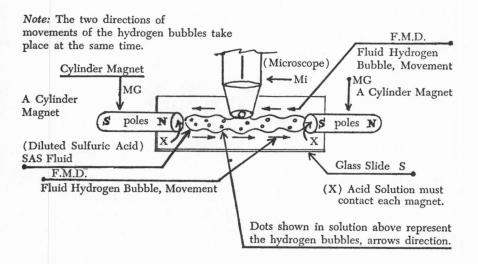

Note: The two directions of movements of the hydrogen bubbles take place at the same time.

Cylinder Magnet

MG

A Cylinder Magnet

(Diluted Sulfuric Acid) SAS Fluid

F.M.D.
Fluid Hydrogen Bubble, Movement

(Microscope)

← Mi

S   poles N   X

F.M.D.
Fluid Hydrogen Bubble, Movement
MG
A Cylinder Magnet

S poles N

Glass Slide S

(X) Acid Solution must contact each magnet.

Dots shown in solution above represent the hydrogen bubbles, arrows direction.

The general movement of a magnet's fields is expressed in most test materials as "the magnetic energy leaves the S pole and travels to the N pole of the magnet or the earth's magnetic fields, as an example." However, in the above research findings, we have taken a diluted solution of sulfuric acid and distilled water, making a solution having a specific gravity of 1175 degrees for use in a fluid liquid channel between the two poles of two magnets, placed as indicated in the foregoing drawing. Use a magnitude from 500 to 1000 X scope lens power. After a period of four to six minutes, the longer the better, the hydrogen gas bubble activity will start to flow. One to two bubbles can be seen flowing from the S pole to the N pole of the two magnets. A few minutes later, a reverse bubble flow will be seen. These hydrogen bubbles flow from the N pole to the S pole of the two magnets. The two directions of bubble movement can be seen clearly. We use the electrical identification formula to best describe the dual directional flow of energies, I for current and E for voltage.

The flow direction and first movement of the hydrogen bubbles described above indicate the similar behavior of voltage in electrical systems. As voltage leads the first action to a flow of electricity, then current lagging behind the voltage starts to flow to support the amount of current that is demanded of that circuit's load. The lagging time of current to fill the demand of the resistance of the load applied is only a fraction of a second's time. However, in the experiment described there is an extension of time necessary to see this flow demand take place, the magneto magnetic current.

The direct and similar law of "The Ohms Law" seems again to be supported in the energy supplied by a magnet. This should also be considered possible in the earth's magneto magnetic flow between the two poles. It is our objective to present new and more acceptable theories of how we may better use the magneto magnetic energies of magnetism. As we point out what appear to be major errors in modern textbooks, we hope these outlines and our discoveries that are a direct result of our laboratory research will result in new laws and concepts and theories that will lead to even further discoveries in and of magneto magnetic energies, a magnet's energies and other electromagnetic forms of energies.

To aid in the understanding of the discovery that a magnet's magnetic field is similar, if not exactly like, to the electrical voltage and current flow laws, we present in part the fundamental Ohms Law to show the principles of magneto magnetic vs. electrical laws:

simple mathematical relationship between the e. m. f. applied to a conducting circuit having a certain resistance, and the current which would flow in the circuit. This relationship is now known as *Ohm's Law*. The law is stated thus: *The intensity of current in any circuit is equal to the electromotive force divided by the resistance of the circuit.*

Expressed in the common electrical abbreviation this law becomes:

$$I = \frac{E}{R} \qquad (1)$$

where I = current in amperes.

E = e. m. f. in volts.

R = resistance in ohms

Equation (1) enables us to calculate the current (I) which will flow when an e. m. f. (E) is applied to a circuit having a resistance (R).

Example: What current will flow through the filament of a vacuum tube having a resistance of 20 ohms, when an e. m. f. of 5 volts is applied?

Solution: The current in a circuit may be calculated by Ohm's Law using the equation $I = \frac{E}{R}$. By substituting 5 for E and 20 for R we obtain $I = \frac{5}{20} = \frac{1}{4} = 0.25$ Amp. Ans.

To find how much pressure or e. m. f. must be applied to a circuit to make a given current flow through a conductor having a known resistance, equation (1) can be put in more convenient form by simple mathematical transformation.

$$\text{Thus since } I = \frac{E}{R}, \text{ then } E = I \times R \qquad (2)$$

Example: The resistance of the filament of a vacuum tube is 20 ohms, and it requires 0.25 ampers for proper operation. What e. m. f. should be applied to obtain current?

Solution: $E = I \times R$. Since I = 0.25 amp, and R = 20 ohms, the $E = 0.25 \times 20 = 5$ volts. Ans.

When the e. m. f. (E) and the current (I) are known, the resistance R of the circuit may be calculated very easily by placing equation (1) in more convenient form

$$\text{Thus since } I = \frac{E}{R}, \text{ then } R = \frac{E}{I} \qquad (3)$$

> Example: An e. m. f. of 5 volts applied to the filament of a vacuum tube sends a current of 0.25 amperes through it. Calculate the resistance of the filament.

Solution: $R = \dfrac{E}{I} = \dfrac{5}{0.25} = 20$ ohms. Ans.

(4)   From Ohm's Law (equation (1) we have $I = \dfrac{E}{R}$

Substituting this value of I, for I in the power equation (4), we obtain:

$$W = E \times I = E \times \frac{E}{R} = \frac{E^2}{R} \qquad (5)$$

This gives an expression for the electrical power in terms of the voltage and resistance.

From equation (2) we have $E = I \times R$.

Substituting this value of E, for E in the power equation (4), we obtain:

$$W = E \times I = I \times R \times I = I^2 R \qquad (6)$$

In comparing a simple one-cell battery and its electrical energy and polarities to a magnet, the first difference is that if we short-circuit a battery all the energy is consumed. If we short-circut a magnet by placing a keeper bar or iron or steel across its poles we only arrest its loss or use of energy. If we do not place a keeper bar across the poles of a magnet it will lose its strength. We are now discussing only a common magnet and not the more advanced rare earth magnets. A magnet's energy consists of polarized atoms of the molecules of the material of which it is made all spinning in the same direction, forming atomic energy in part. However, we must keep in mind that the center of each and every magnet contains a magnetic equator and it is at this point that the atomic electron spin changes its phase relationship and spin to the opposite direction forming the two potentials—one positive (S pole) and the other negative (N pole).

A magnet can lose this polarization energy (magnetism) by allowing the energy to flow into space, losing its energy by atmospheric space absorption, such as the straight line emission of energy that leaves the ends of the poles.

When we apply a magnet to perform a specific function its energies may be drained, such as pulling reeds (metal plates) in the activation of a relay or to lift metals. Each time a magnet's energy is applied a portion of the energy is lost in the metal to which it is applied. Physical contact results in certain transfer of the polarized energy to and into the metal the magnet is applied to. Here we are discussing standard permanent magnets. Their use is in motors and other more sophisticated devices which also act to consume the magnet's energies.

On page 58 we show a test that can be reproduced to show the dual and opposite directional flow of energy that takes place in a magnetic magnet circuit.

To show how to short out a magnet, remove its energies, without A.C. currents being applied to the magnet, the conventional method to remove magnetism from a magnetized metal is to lower the magnet into a glass beaker of sulfuric acid. The magnetic fields will be absorbed in part in the acid. The acid would then act to short out the magnetic energy by speeding up the loss of the magnetic energy by acidic absorption. This test is worthy of reproducing as it shows a curve of gauss loss to the magnet's normal gauss strength prior to placing the magnet in the acid bath. The time curve is hours to days to lower this energy that would normally take years to do under normal magnetic operational uses.

Indications from these tests in our laboratory are that the magnetic energies are transmitted from the ends of the magnet in straight or linear lines, as magnets have a frequency and/or wave formation similar to that of a wave length. Our research has also shown that magnetic energy can be mode-modulated by application of other frequencies. This provides an intermediate frequency result, an I. F. frequency. Again, we see a new approach to the use of magnetic energies including those from a magnet that were previously unknown. This discovery should lead to new discoveries and developments to serve electronics, biochemistry, biophysics, physics, and all fields of applied sciences.

On page 64 we show a chart indicating the many frequencies of energy as the overall frequency spectrum. The frequency spectrum chart is presented for better understanding of the laws of physics for our further discussion. Each form of energy is motion. This motion is in cycles per second of generated energies, energies that combine as to the frequency, the number of cycles of energy that is present, to make possible the division between energies and their types and forms.

In discussing any new scientific concept we have found, in writing or lecturing on the subject, it is better to start at the very basics of that science, which provides a firm foothold for a better understanding of the subject, the improvements to the concept, theory or practical mechanics of the discovery or development in the presentation.

At this point we present another misconception of magnetism. It is today presented in textbooks and the general understanding of a magnet's magnetism that the energy transmitted by a magnet is composed of "Lines of Force."

Our laboratory findings show that energy radiated from the magnet is in fact not lines of force but small cables of force.

THE CABLE EFFECT

Due to the importance of properly knowing and understanding the makeup of magnetic energy coming from a magnet, we continue here to present the findings of our laboratory over many years of development. Another finding is actually seeing in part that energy that is transmitted from the poles of a magnet. It is possible to obtain photographic pictorial outlines that allow us to see the magnet's energies as they in turn affect the scanned 400 apx lines of electron sweep appearing on the internal face of a color dot cathode ray tube. Bringing one end of a magnet to and against the exterior glass surface acts to cause the energies from the magnet to displace the horizontal scan and vertical scan lines on the tube's surface. This activates the color dots in an outline of the applied energy from the magnet's poles. This resultant displacement display can then be photographed in color by the use of color film with a F 1.8 lens on a good, well-mounted camera with a time exposure of 11 seconds or less in a totally dark room. The results are not lines of force being emitted from the poles of a magnet but are miniature-size cables. These cables are several thousandths of an inch in diameter at the very end of the magnet pole. As they travel to 1/16 of an inch to many inches, these

Outline Representing in Part the
Frequencies, Number of Cycles per Second Found
in the Audio and Electromagnetic Spectrum

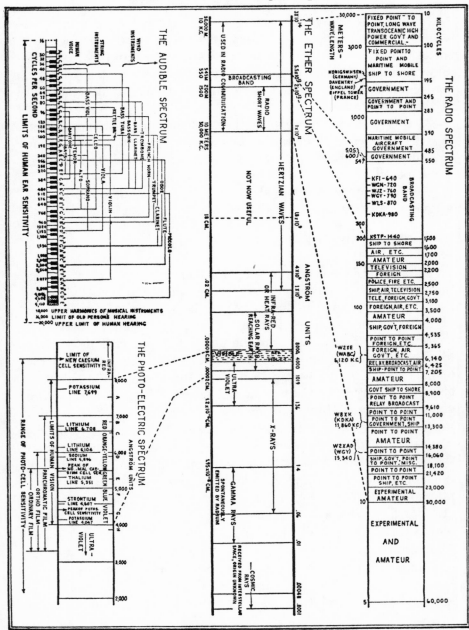

miniature cables enlarge and allow a view into the interior of these cables. The spin effect is also noticeable by a pull, an electronic vortex twist, that appears on the screen at the outside edge of each cable and/or the roster. The horizontal scanning lines that appear on the color dot cathode ray screen are pulled in the direction of the magnet's energy electron spin. The N pole acts to present a left-hand spin in relation to the pole position and that of its directed position to the screen's surface. The S pole then acts on applications to the surface of the tube to present a right-hand spin. The cables also take on and present this electron spin effect. The center of each cable contains an energy that is opposite to the outside electron potential form. There is much more to this cable effect and we hope at a later date to make available a book on the discovery of the twin energies of the cable effect on magnetic magnet energy.

Many theories and concepts persist in new methods to protect men in space from harmful radiations of magnetic fields. One particular concept is illustrated in the accompanying drawing.

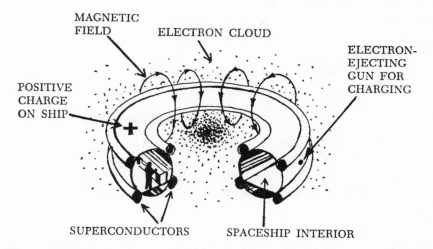

MAGNETIC FIELD   ELECTRON CLOUD

ELECTRON-EJECTING GUN FOR CHARGING

POSITIVE CHARGE ON SHIP

SUPERCONDUCTORS   SPACESHIP INTERIOR

"Plasma shielding," an alternate plan, puts positive electric charge on craft to repel protons—and uses superconductive magnet's field to prevent approach of electrons that would destroy charge.

In the presentation the use of super conductive tape wound magnets is suggested. The tape is wound to make electromagnets having no resistance to the flow of electricity by supercooling. The belief was expressed that a very high magnetic charge would shield persons inside the container against space radiations.

Although this concept was and is still novel and instructive, it would not protect persons in the space container from the effects of the magnetic fields due to the construction.

Our laboratory findings are definite on the biological effects magnetic fields have on living systems. The high magnetic field environment created in the suggested design have dangerous side effects on the humans in the container.

From our research findings we offer the suggestion that man in space can be better protected from space radiation in an environment suitable to earth's environment, and man would not face a magnetic field lower in space than on earth. It is known from space technology that leaving the earth's magnetic environment to a lower magnetic environment can result in serious discomfort or death from combined exposure to the lower magnetic energy.

Our solution, based on our research findings, is to design suitable magnetically treated clothing or a wall-to-wall internal magnetic environment of the ship's inner surface that will provide the space travelers with protection against lower than normal magnetic environments. Also, the outer skin of the ship may be provided with an electromagnetic charge that would assist in offering protection against many types of space radiation. This can be done so there would be no effect to radio communications or equipment operations. The electromagnetic charge can be applied within the ship or located outside the ship.

The effects of lower than earth's magnetic field to living systems are well known to NASA's space medical directors; they are deadly and most harmful. As we write this we believe that from the developments we have made there may come a partial answer to many of the problems facing future space research undertakings. These developments show the direct possibility also that magnetically treated clothing can be used to upgrade strength. Special treatment of the clothing can also aid in the recovery from atomic fallout exposure accidents that result from handling, using, or accidental exposure to atomic energy and/or X-ray radiations.

Our discoveries show that when certain magnet magnetic energies are used in adaptable clothing, arrest and recovery are practicable in a great number of cases from the harmful radiations as described.

Chapter Eight

WHY PREVIOUS EXPERIMENTS
HAVE FAILED TO BE REPRODUCIBLE

Since the late 1800s research has been carried on in many countries of the world on the effects of a magnet's energies. While a great deal of important research data and resultant papers that were written are still presented on the biological effects of magnetic fields, much of the early work as well as the recent work is not reproducible to obtain definite results.

In this chapter we will present some facts as to why these experiments, many of them of great importance had they been reproducible, could not be reproduced time after time with the same results.

Figure A shows a laboratory horn-type magnet; its poles are indicated by N for north and S for south. A microscope slide is shown between the two poles. Note that the subject matter B on the slide is nearer the N pole than the S pole. By physical placement nearer the N pole a greater distance exists between the subject matter and the S pole of this magnet.

Figure A   Figure B   Figure C

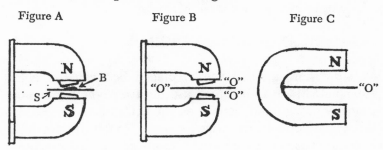

N   is North Pole.   S   is South Pole.   Slide marked S, and Specimen B.

The culture or biological specimen then would not be properly aligned to receive both pole effects. It would, in fact, be receiving

67

more of the negative pole energy than the south pole, which is positive energy, and part of the specimen would be in the center of the Bloch Wall effect separation. This separation is where the S pole energies as they travel toward the N pole would be in the center of the 180 degree phase relation change that takes place between the poles of each and every magnet.

Therefore, when this experiment was duplicated there could be a change in the physical positioning of the slide and its specimen to the pole effects. The results then could never be exactly reproduced, and should it be that it was possible to get this direct exactness in positioning, the effects would be in three parts, not one. To explain this refer to figure B. We see the exact center of this horn-type magnet's poles indicated by a straight line. This is the center of the 180 degree phase change. This line indicated by the letter O further indicates that we have no magnetic lines of force that carry either the negative or positive pole effect at this position. Instead, we have a division or zero point of energy radiation. The same exact occurrence would take place if we used a horseshoe magnet—see figure C. The line of separation is again marked with the letter O.

Should we place a test tube of fluid upright between the two poles of a magnet, we would, if its positioning was exact, obtain the dual effect energies and the zero magnetic potential effect as indicated by the phase change point O. We can see we have in fact a three-stage potential radiated effect to the specimen, whatever the nature of the specimen might be.

SMALL ANIMAL RESEARCH

We show here a small animal placed between the poles of a large magnet. Many researchers use electromagnets—magnets having exact windings and connected to D.C. power sources. There is a difference between the effects one can obtain from a solid state metal or composition magnet and those of an electromagnet. We discussed this earlier as to magnetic wave frequencies. In the poles shown, N is north pole and S is south pole.

The popular positioning of small animals or rodents between the poles of magnets is to place the animal in a confining cage made of aluminum wire mesh, positioning the animal lengthwise

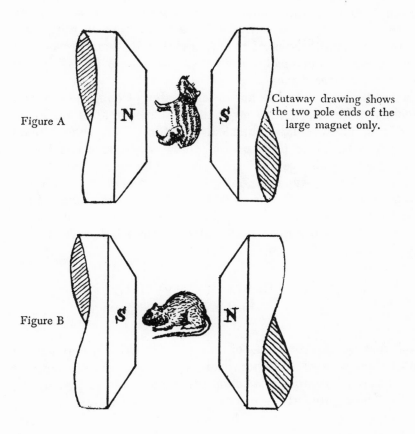

Figure A

Cutaway drawing shows the two pole ends of the large magnet only.

Figure B

between the poles. In this manner half of the animal's body receives the S pole energies and the other half the N pole energies.

We show in these drawings an animal lying sideways. This was done to show that each end of the animal is receiving only that particular pole's energies, and the middle of the animal is receiving the O Bloch wall magnetic field separation of the zero magnetic field, the energy result of the center of the 180 degree phase change of the magnetic spin effect.

In scientific books written by authorities today and those by researchers who conducted similar research many years ago, we find the ironclad theory that both poles present a "homogeneous field." While they refuse to change their thinking toward the facts we have attempted time and again to present to some of these scientists, their work will continue to be based only on misguided and mistaken applications of magnetic fields in their research of biological and biochemical systems. The two pole fields are not homogeneous or the same.

Each pole has a separate and totally different effect to any and all specimens to which it is applied. This encompasses all living matter—airs, gases, solids, chemical reactions. This should open new doors to advanced physics using this newly discovered law as a foundation. While we say new, your senior author discovered this law in 1936; however, scientists and researchers have refused to accept this law in the past. Now is the time to update their research and allow a better approach to advance this vital science since it is a science important to all the known sciences of man today.

MUTANTS ARE DEVELOPED BY THE
OLD CONCEPT APPLICATIONS

Let us now take the case of the growth and development of mutants. One species that has been widely used in biomagnetic research is the *Drosophila melanogaster*, commonly known as the housefly. The continued use of equipment to hold the young fly or the larva or the eggs prior to hatching has resulted in and after exposure to the magnetic fields of a magnet in producing, as growth and development proceed, mutants. These experiments have produced specimens having one large eye, one withered wing, two long legs, one short leg and one long leg, and so on. This is magnetic mutation of the living system by controlled radiation of the fly and can apply to insects, animals, or living systems by selected magnetic fields.

The side of the insect placed next to the S pole will grow at a faster rate since the S pole energies are the production of the positive right-hand electron spin which causes an advanced growth and development of any and all biological systems exposed to it in the early stages of development before maturity has been attained or by treating the larva before birth.

The reverse takes place when half of the animal, insect or

other living system is exposed to the N pole for suitable periods. The strength of the magnet used depends again on the technical researcher as to the subject of his research. We have found that from 1,000 to 4,500 gauss is the best curve or range from low to high gauss that effects the best overall results of alternation to the living system.

It is possible to program the changes of the genetic attitudes of animals, plants, insects. We can change the normal genetic growth and development to abnormal products of living systems so planned and computed by advance knowledge and planning.

Total exposure would then result in the living system being placed in either the S pole for advancement of the genetic programming or the N pole for arrest or development of the genetic size and health of the system.

We emphasize to the student or professional researcher that even greater changes can be found by even further and equal discoveries of the effects from a magnet's energies using other forms of concept discoveries. This is a vast field in itself and should properly be presented in later releases.

The genetic changes we have programmed in our laboratory experiments and findings were important in other research work on cancers and cancer tumors of many types and stages of development.

THE POSSIBILITY OF DIAGNOSING HUMAN ILLNESSES WITH CONTROLLED MAGNETIC FIELDS

From the work and research we have conducted with animals having a blood type and circulation system similar to man, we foresee in the near future the release of a system for the detection of nearly all human illnesses by application of the fields generated and transmitted from a magnet—solid state or electromagnet.

The disclosure is based on the fact that it is now possible to screen many complaints in animals by the physical reactions presented by the animal's nervous system when select and controlled magnetic energies are applied to affected parts of that system. This discovery was made while working with the nonhomogeneous magnetic fields on animals and measuring their physical responses under the separate and applied fields.

Chapter Nine

CANCERS AND TUMORS
AND MAGNETISM

In order for this book to be read and understood by the layman, student and professional researcher, we are refraining from the use of highly technical material or language and the display of mathematical computations. These materials are, however, available for proper usage.

Over 300 active cancer biopsy transplants to laboratory strain white rats, mice and rabbits, each having a similar blood type to that of man, 89.6 percent, were programmed as to growth, development and/or arrest of the cancer. This was accomplished through the proper application of the arresting energy of a magnet's poles. This was the N pole or negative electron spin effect of that pole's energies. Years of research found that one effect of cancer development is a genetic transfer from one generation to another, not necessarily in that order. A genetic carryover from one generation may not be found to become active until the third or fourth generation. Yet, positive information as a result of some 18 years of these studies is convincing in our findings in this area of study. During the 18-year period, two years were completely devoted to these studies. Cell genetics can carry the active cancer seed for future generations and development once the cancer seed is active in a living system. This does not apply to all cancers. There are types that are local and can be arrested and do not carry over as genetic transfer.

We will not entail a detailed discussion on what causes cancers since there are as many causes of cancers as there are types and kinds.

In our earlier discussions on the electromagnetic effects to mankind we discussed the outer space electromagnetic effects on man and his biosphere (biological atmosphere). Several causes for cancer were presented in that discussion.

At this time in our history there are over 100 types and kinds of cancers. What may arrest one type will not necessarily have an affect in arresting other types. Clearly we can see from such facts that no one agent will arrest all cancers. There has been no agent, process, drug, or development to combat all cancers effectively as there is no known single cause for the development of cancers.

Accepting these scientific facts as we know them to be, should we find a single form of combating agent that would arrest all cancers this would be a remarkable and outstanding discovery.

There are various forms of treatment that will arrest cancer development of many kinds and types if they are caught in their early stages of development.

We are interested in the early stages of development and equally in the advanced stages. In each phase we find that when magnetic energy of the negative N pole is applied to the cancer site, a remarkable reduction in the condition and also a marked arrest in further development of the cancer condition takes place.

Our laboratory has not at this time been able to secure biopsies of active cancers of all of the more than one hundred types known to exist. We have, however, obtained through medical doctors and professional people engaged in cancer research many different samples of active cancer tissues for transplant and development. It is well known that you cannot transplant any form of cancer and have it take unless there is a state of infection at the transplant site. Cancers will not take when transplanted to healthy tissue. In our laboratory we have infected that area planned for active transplant of cancer tissue cells prior to making the transplant. The infection has failed to take in some cases and had to be repeated, but it will infect eventually unless there are adverse conditions to this taking or acceptance action on the part of the research animal. Prior magnetic exposure has prevented an active take of infection even after several attempts have been made to graft an active cancer transplant. A form of antiserum, or defense mechanism, seems to have developed in a number of cases to prevent the cancers from taking and developing after such exposures.

Prior to the active cancer cell tissue transplants the animals were radiated, exposed, to the negative energy of the N pole energies for several hours. This acted to resist our attempts for a successful cancer graft to result or take. Take is used here to mean the infected tissue accepts the active cancer transplant and develops into a state of multiplying cancer cells.

The encouraging fact is that when tissue is exposed to negative magnetic energies prior to the transplant of the cancer-infected tissue, there is a noticeable resistance to its successful development even to previous infected and prepared sites.

Here we see "What is now needed is a new approach as to the unification of energy." This statement is not original. In fact, a number of outstanding researchers have made this statement within the past few years. It is becoming more apparent that unification of the findings in the many fields of applied cancer research is needed to obtain a fresh, new, theory or concept for use of all the sciences that have shown good results in arresting the development of cancers.

Should we start at the very beginning of the structure and basic energy of all biological living systems which makes it possible for them to live and multiply, we would find the atom and its electrical system, which is a great teacher for new approaches and better understanding of what we are working with and our purposes.

In making in-depth examinations of the applications of electrical and radiological energies which have proven to be partially successful in arresting all stages of cancers, we should study the laws of the atom—the electrical building block of all matter, including all biological life forms. From this study of the atom we can see it is possible to arrest any form in the living system, including bacteria, malfunctions, and mutations of cells, tissues, organs and glands. Each of the aforementioned segments are the result of molecules being built from single groupings of atoms.

When this primary and most elementary law of physics is applied to living cells, we see a well-balanced electrical system, and any changes will upset the well-regulated bioelectrical system of the cell, causing it to deform, mutate, or break down. Since it is now possible to measure and record the voltage existing on the outer surface of the blood cell's membrane, the first recordable signals tell us there are difficulties arising in the structure of the membrane supporting walls of the cells; as in the case of the red blood cell we find a rise in the negative voltage. The rise in negative ion charges on the outside of the membrane wall is then compared with what happens to the atom. When there is a rise in the negative electrons on the outer vortex of the atom, the atom is no longer the simple type of atom that makes up a substance; it is then altered to make the molecules form a more complex structure of the element or substance. We might then compare the simple hydrogen atom and the second type of the same hydrogen complex that is slightly more complex, yet, nevertheless, remains a hydrogen atomic complex.

The blood cells' bioelectrical ion charges, which result from the charges taken on in the form of sodium and potassium ions, their charges, and level of charges, depend on the selectivity of the walls of the membrane.

Anything that changes the selectivity or the charges of the cells and their membrane supporting structure will affect the health and welfare of the cell proper. Other direct research findings that link the membrane variations and the transformation of normal cells into malignant cells present a linear curve as to the voltage measured across the normal cell, and the rise above normal negative voltage found to exist on the cross axis measurement of the malignant cells. In further investigations, we find in all cases of human or animal biological stages of internal repair we have a rise in the negative potential on the outer surface of the affected section. This also follows a resultant linear curve as to the amount of negative voltage potential as to the degree of damage compared to the rate of natural healing of the affected part to the living system.

However, here we depart from the normal. The cells that develop into malignant cells at first show a rise in negative potential voltage across the cell's structure, a slow change takes place, and we find when the cells are fully developed as malignant cancer cells the negative voltage across the cells drops to a lower than normal negative voltage potential.

The first effect is the rise in negative voltage which we find happens in all damaged segments of the living system. The resultant drop in below normal negative voltage across the affected segment tells us that repairs have not been made and that part, segment or cell is a mutant no longer under the repair control of the living systems' defense mechanism. This is noted by the negative voltage being lower than normal. We note here that normal means the normal potential found to exist on the same segments, and cells that are present when there is no difficulty experienced in or by that segment or cell's normal operating potential.

In further study of this effect we find nondividing cells have a high normal potential negative voltage opposed to a very low or lower than normal voltage existing on the cross section of dividing cells. Dividing cells means cells that are affected and are malignant, including rapidly proliferating tumor cells.

As a direct result of these galvanic potential related energies, we find whenever we have damage to any part, segment or cell of the living system we have a higher than normal rise in the negative voltage potential. When normal healing and recovery results the negative returns to normal. If we then find a lower than normal negative to continue we have a section that has failed to recover and return to a normal state of health.

On page 78, Item A shows a normal red blood cell. Item B represents a form of malignant cell—note the deformed membrane outlines in Item B. Item C is an electron microscope photograph reproduction of a healthy red blood cell. Item D is a red blood cell that has decomposed, showing the breakdown and resultant misshaped cell when affected by the presence of a distorted potential of biological voltages. Migration by active electrolytic transfer from one cell to the next results in any number of associated blood decomposure forms of diseases. In every case of advanced malignancy there is found to exist a lower than normal voltage across the cell's membrane and an increase in the plasma and/or fluid amounts and flow between each and every cell. To present this rise in fluid flow and the distortion of the cells membranes, we show on page 79 several low microscopic magnification photographs ranging from 60 to 100 X power of blood cell patterns. These are a result of drawing a few drops of whole blood from a fingertip, placing on a microscope slide, and allowing to dry, and then examining them under the low power magnification. Using this most elementary method of screening the blood, a number of very interesting facts are brought to light.

No. 1 on page 79 shows a microscopic enlargement taken by a 1800 power microscope of a few normal blood cells. We use this as an example.

No. 2 shows the increase of plasma, fluid, that can clearly be

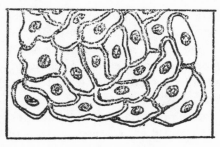

ITEM A

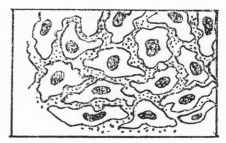

ITEM B

ITEM C

ITEM D

seen between the dried blood cell clustering. This sample was taken from a male, 44 years of age, suffering from cancer of the supermaxillary with metastasis. Study here shows profound disturbance of the clot retraction pattern.

No. 3 shows a blood sample of normal blood after drying and using the low power 60 to 100 power microscope lens for this screening study.

No. 4 shows a dried blood sample of a male, 70 years of age, suffering from advanced leukemia. Note the almost identical pattern to that of one of the many and foremost types of cancer.

No. 5 shows a pattern of dried blood of a 12-year-old female, symptomless. Note normal blood clotting pattern.

No. 6 shows a dried blood sample of a female, 34 years of age, four months pregnant with no cancer development. Note the similar pattern of the disturbances of the blood and the increase of plasma, fluids, separating the blood cells and blood cell clusters.

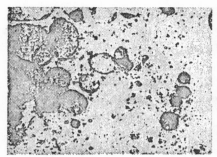

No. 1

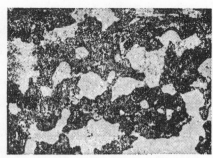

No. 2. Male; 44 years; cancer of the supermaxillary with metastasis; profound disturbance of the clot retraction pattern.

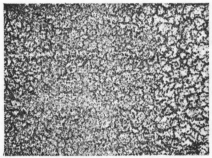

No. 3. Male; 55 years; symptomless; normal clot retraction pattern.

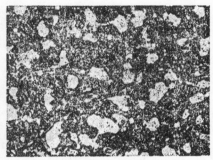

No. 4. Male; 70 years; leukemia; disturbance of the clot retraction pattern almost identical with cancer.

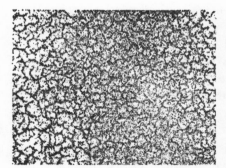

No. 5. Female; 12 years; symptomless; normal clot retraction pattern.

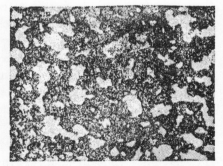

No. 6. Female; 34 years; but four months pregnant; advanced disturbance of the clot retraction pattern.

These microphotographs show us in this simple form of screening the changes in the amount of fluids, plasmas, that flow between blood cells clusters when there is an abnormal condition existing in the system. Abnormal meaning not as we would find blood serum flowing under normal conditions. These photographs also show how pregnancy alters the amount of fluid serum flow between the cells and that which happens when cancer is present.

Photograph No. 6 presents the clue that when the body is under stress as in the condition of pregnancy there is a weakening of the normals as the embryo in forming is calling on the human system for blood and all the basics needed to support this additional life developing in the body and the system devoted to new human development. This is an overall demand which acts to upset the serum balances of the cells while not the cells themselves.

The cells in the case of changes in the blood serums and/or fluids are not affected as they are in stages of cancer developments. In cancer developments we have a destructive mutation of the cells' membranes.

In the case of cancer, the increase in the separation of blood cells and the clustering separated by the serum amounts, as an increase in the amount of blood fluids, plasma, this fluid then acts to be a carrier for the escaping electrolyte that flows from the malignant cell and it now appears to act to upset the electrolyte fluids of other cells. We then see the possibility that this effected electrolyte can and may be the carrier or active transport for fluids escaping from the decomposed cell walls of effected cancer cells.

There are two major theories which are foremost today in research into the causes and arrest of cancer. One theory is that a virus may be responsible for the start of the condition. The second theory is that food intake, chemicals in foods, or for a number of reasons, a condition develops causing a disturbance in the blood cells' balance as to the biological chemistry of the cells, therefore, the tissues.

Today researchers know more about cancer development than at any other time in history. However, the mystery of why cancer develops in the human or living system remains. We have discovered in part that tumor cells develop and multiply rapidly without any apparent control responses. Further, they have the ability to spread rapidly throughout the living system. It has also been discovered that natural sunlight can cause cancer of the skin, as can other energy radiations. Chemicals and gases, such as cigarette smoke, can trigger cancer of the lungs. Certain viruses when transplanted into laboratory animals can also cause cancers. On eating the flesh of livestock that have been fed certain feeds containing a number of chemical growth stimulants, cancer again has been promoted. There are a number of viruses that when injected into rats, mice or rabbits develop into cancers and are transmitted by injecting the blood from one to another. If we study these transferable reactions that act to trigger the cancer cell development, we see there is a good chance cancer is caused by a delayed to prompt response of the virus.

THE EFFECTS OF MAGNETIC FIELDS ON CANCERS

Of the several hundred research transplants of cancer to rats, rabbits, mice, and other animals, it has been proven that the N pole, the magnetic negative energy of the two poles and their separate energies, has slowed, controlled and arrested further development of the active cancer site. Better than 90 percent of the cases so treated have shown a control and arrest of the cancerous condition, depending on the state of advancement of the cancer and the age and physical condition of the animal in question.

To further support this finding, when the S pole of a magnet, this being the positive energy of a magnet, is applied to cancers they become more advanced and then develop, grow and spread at an accelerated rate. Therefore, the two effects prove that the energies generated by a magnet's two poles can and will prove in the future to be a new application of a very old science toward the arrest and/or containment of cancer development.

As pointed out earlier, when any disorder, break, certain infections, or physically damaged parts of the living system are placed in a strong negative energy field there is almost at once an arrest of further damaging developments. We also discussed earlier that nature itself directs a negative electrical field to a broken bone or other damaged segment of the living system under attack. We feel this shows in part the effects of controlled negative energy effects of the north magnetic pole to aid in the arrest and more rapid healing of many, if not all, forms of disorders to the living system.

THE CONTROL OR ARREST OF
CANCERS BY MAGNETO THERAPY

Leaving the effects of magnetic fields on cancers and tumors, let us review the findings of the work in our laboratory to date and the results we have obtained by the use of this energy on actual cancer transplants, which have shown that we have a new and vital tool to combat, control or arrest many, if not all, types of cancerous conditions.

If the cancer is deep beneath the skin's surface, care must be taken to insure that the cancer site will receive negative (N pole) magneto magnetic energies having gauss strength from 2,500 to 4,500 and that this amount of energy is not merely applied to the outer flesh. To assure proper dosage to the cancer site, calculations must be made by measuring in inches the distance from the N pole of the magnet to the depth the researcher wishes the treatment exposure to reach, then selecting the proper magnet or electromagnetic power having those certain gauss energies at that distance from the pole end. This can be done by taking a magnetometer reading at X number of inches from the N pole of the magnet selected. Naturally, should the cancer be within the body, this means we would have a higher than 2,500 to 4,500 gauss strength on the body's surface.

THE INCREASE IN THE PRODUCTION
OF ERYTHROCYTES

Using the results of over 300 cancer research test cases of animals having a blood type similar to man, the findings of this work show:

An increase in the production of erythrocytes in peripheral blood while proliferation of leukocytes is inhibited as a result of exposure of the cancer site to the negative N pole energy. These findings are documented by transplants of cancer, A through C classifications, to large and small rodents and to other animals.

These transplants having developed into first to third degree types, exposure thereof to the negative N pole energies resulted in an average of 88 percent arrest of further development and 87.5 percent recovery. The number of cases tested was 290 to 325. Success of this project was, in part, due to careful handling and exact and proper exposure over several months. Again, the fact remains that any exposure of the cancer to the positive S pole energies immediately caused an advancement of the condition. The S pole energies are positive and are similar in part to the effects of certain radiological energies being used today in an attempt to arrest cancers. Additionally, we have discovered that the negative energies of the N pole of the magnet also strengthens the unaffected surrounding cells, generating what appears to be an increase of the natural defense mechanism of those unaffected cells to combat spreading of the cancer.

In reviewing our research work with electromagnetic energies, we find we are working with an energy closely related to the one provided by the living system to the cells and tissues of the body which acts as a natural barrier to any form of cell or tissue breakdown. The results of our research work have shown that application of this natural energy may open doors for additional research and development not only for the control of cancer conditions but for a new approach to control, arrest and prevention of many of the diseases medical sicence now finds difficult to cope with.

In studying the two immunological defenders produced by the living system, these being B lymphocytes, which are manufactured by the bone marrow, and T lymphocytes, which are produced by the thymus located at the base of the throat, these two important fluids, which are antibodies, present the body with a natural defense mechanism to arrest, attack and control the numerous invading viruses,bacteria, infections, etc., that the body must defend itself against.

Although many drugs have been developed today, nearly all of these drugs result in poisoning of the cancer site or total or partial poisoning of the living system. Many researchers believe these drugs are the answer to arrrest of further spreading of the cancers. However, poisoning of the system with alkaloid pharmaceuticals is only partially effective and is extremely harmful to the body. Yet, if the body can accept this toxic matter, some improvements will be shown but at the expense of the overall health of the patient. Therefore, we are convinced that attention should be directed toward the use of negative N pole energies and its applications for the arrest of cancer in either its primary or advanced stages, thereby giving the patient an advanced method of relief or the actual arrest of the cancerous condition.

From our years of laboratory research into the causes and arrest or relief of cancers by the use of magneto magnetic negative energies, we fully believe cancers are a direct result of an internal virus existing in the body or are actively transmitted to the body by external biological and/or chemical atmospheres, or a combination of these factors. Although not apparent during the life span of most men and women, we feel a cancer virus exists in all living systems and under ideal conditions may be triggered into a cancerous state. Evidence supporting this theory has been reported by other researchers. With the research now being conducted throughout the world, we believe an announcement of an anticancer vaccine could be imminent.

However, until an anticancer serum which will provide resistance to this disease is discovered and developed, the development and application of negative magneto magnetic energies could save millions of lives annually, worldwide.

The use of X rays, cobalt and radioactive energies are all of value. Again, we find that in each case of their use there is resultant poisoning of the living system. Within the composite electromagnetic formula of vibrations of atomic energy that these energy forms have and transmit to the living system, there are positive ions' reflective energies which in nearly all, if not all, cases act to advance the body's acceptance of these positive energies, which promotes many phases of the cancer condition. The entire projection of desired results is in the fact that these forms of energies will sometimes arrest certain types of cancers. If we were to agree with this accepted method of cancer control, we would need to see a considerably higher percentage of arrest than is now indicated. Here, as in the pharmaceutical drug usage, we find the cells adjoining those infected with cancer are adversely affected, lowering their defense against the continued outer wall membrane breakdown that is a result of cancer and is also a result of radiological exposure treatments.

Our investigations show that by the use of negative magneto magnetic energies we arrest the breakdown of adjoining cells and also arrest further progression of the cancer condition. We also strengthen the adjoining cells to act as a natural defense against further cancer development.

Further, we have found nerve pains and physical pressure resulting in nerve pain can be greatly arrested by the application outlined above used to arrest and control many forms of cancer development.

Chapter Ten

THE BIOELECTRICAL CONTROL OF NERVE PAIN

As in the accepted theories and findings of nerve bioelectrical activities, we find the nerve's bioelectrical systems entertain exactly reverse galvanic energies to the measurements and findings of those existing in the blood and tissue cells of the living system.

Blood and tissue cells have a negatively charged sodium exterior membrane and a positively charged potassium interior. The outer fiber covering of the nerves has a positive charged sodium ion condition and a negative interior potassium charge, which is the exact reverse of the galvanic potentials found within the blood and tissue cells.

When nerve endings are affected by abnormal pressure, infection, disease, or a severed condition, they exert a potential energy that automatically informs the brain of danger, damage or pressure —internal or external. When we apply magnetic negative energies to this affected nerve state, there is a lowering of the positive external potential of the outer nerve fiber coating, resulting in a sedated action. This is caused by the lowering of the sensitivity of the nerve since its highly effective and sensitive positive ion potential has been reduced. The inversion of the positive ion charge existing on the nerve's surface is directed to the N pole magnetic energy field, which lowers the nerve's sensitivity and its positive ion sodium charge, resulting in less galvanic transfer of pain information reaching the brain for translation.

Shown on page 88 are drawings of the nerves and a brief review of their working order. These drawings are included for better understanding of the effects of negative energies of the magneto magnetic emissions on the nerves. Drawing B illustrates the power plant of the nerves, and Drawing A shows how this energy is transmitted to the muscles.

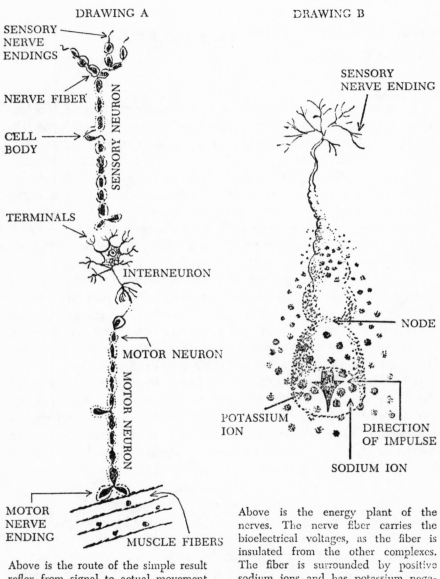

DRAWING A

SENSORY
NERVE
ENDINGS

NERVE FIBER

CELL
BODY

SENSORY NEURON

TERMINALS

INTERNEURON

MOTOR NEURON

MOTOR NEURON

MOTOR
NERVE
ENDING

MUSCLE FIBERS

DRAWING B

SENSORY
NERVE ENDING

NODE

POTASSIUM
ION

DIRECTION
OF IMPULSE

SODIUM ION

Above is the route of the simple result reflex from signal to actual movement of the muscle.

Above is the energy plant of the nerves. The nerve fiber carries the bioelectrical voltages, as the fiber is insulated from the other complexes. The fiber is surrounded by positive sodium ions and has potassium negatively charged ions in its center.

The effects of applying N pole magnetic energy to the nerves act to lower their sensitivity. This lowering of sensitivity allows us a certain control of a pain condition. When we transmit S pole energies to the nerves they respond with a greater sensitivity to pain. We then have the reactions to the two separate pole energies of the magnet's negative and positive resultant effects.

Referring to Drawing B, the generation of energy within the nerve itself results when a sensory nerve ending is stimulated. The sodium ions are quickly admitted into the nerve at the node, this being a break in the insulation. The following reaction sets off a chain of pulse charges that extend from one node to the next and is carried to the center of the nervous system.

Referring to Drawing A, the energy from a nerve reflex, these being impulses, presents a stimulus that ends in the contraction of the motive muscle. This energy transfer is made possible by the neuron as it triggers this energy which flows along the nerve fiber through the cell body to an end in the terminals in the spinal cord. At this point the interneuron acts to transmit these impulses to the motor neuron where these impulses act to motor the muscle.

It is equally important we consider the fact that when there is no disease present, the use and application of the S pole energies can then be used for strengthening the nerves and their resultant responses. This offers still another possible new tool for medical researchers in restoring the nerves to normal or near normal activities in cases where such encouragement is needed.

Not only the nerves but the muscles also may be stimulated, if and when needed. The offerings that the S pole energies can deliver when applied to nerves and muscles also applies to all organs, glands, and segments of the living system.

The speed of the heart can be controlled, as well as its duties. Considering these values we find ourselves with a new and important tool to aid mankind find the long sought answers to many complaints which for generations have caused great concern. This also encompasses rodents, mammals, and all forms of life.

Since it has been found possible to stimulate, as well as control and regulate, glands, cells, and tissues, we then consider that these discoveries may encompass aid to patients suffering from many types of mental illness. Help to these persons could be considered since all food taken into the system acts in many ways to affect mental attitudes and the resulting reactions. Take as an example the amino acids. Typtophan is one of the some 20 amino acids that are the building blocks of proteins which in turn manufacture serotonin. Serotonin, in part, is responsible for the means by which neurons carry messages to parts of the brain—it is an active means of information relay transfer. The loss of serotonin can activate an oversexual interest. A normal to excessive amount of this serum acts to encourage mental processes directed to the normal human behavior patterns, providing no other condition affecting mental behavior exists.

THE AMINO ACIDS AND RESULTANT PROTEIN DEVELOPMENT

Taking a sample of any of the many types of amino acids—the basic building block of the protein structure—and exposing them to the S pole energies of a biomagnet, can and will inspire a higher degree of energy and development resulting in a higher valued substance. These facts were discovered in the research toward the arrest of cancers by dual and separate magnetic pole field effects.

Everything that can be done with the positive S pole energies of a magnet to inspire strength and biological developments can, to a large degree, be reversed with the negative N pole energies. This was discovered after hundreds of exposures of these energies to the living systems' glands, organs, or segments, as each responded in the same manner.

The facts, as they now present themselves, are that when we use the N pole negative energies we do not stop the production or development of life systems, instead we simply arrest their actions or developments without the use of toxins or poisons to obtain this effect. This provides us with a far better approach to disease arrest, as in the case of cancer treatment. The methods presently used cause a radiological poisoning of the system, in addition to the effects to the cancer or tumor site. We know within the scientific community this may be subject to some arguments. However, the facts speak for themselves, and the overall improvements the science of biomagnetics offers will outweigh the objections of the few who may oppose its use. We feel if these objections should occur they would be based on the fact that those who make them do not have the practical work and, therefore, have as yet not obtained the full impact of this science and valuable approach to this work.

Chapter Eleven

MAGNETISM AND GRAVITY

Earlier we discussed the reactions magnetism, electromagnetic energies, and atmospheric energies have on man and the biological system. This also includes plants, which are living systems.

While we have up to this point discussed these energies and their effects on living systems, we will now show how magnetism and gravity are combined and react together, as we cannot have one without the other, and how these two energies affect all living systems.

We have shown that a magnet is composed of two forms of energy that are similar yet different in their energy charges and potential. If we take two long straight or cylinder magnets, then place the south pole of one magnet in contact with the north pole of the other magnet, they are no longer two magnets but have combined to become one magnet. There is only one north pole at one end of the magnet and one south pole at the other end of the magnet. Where they came together we find the Bloch Wall, the point of no recordable magnetism.

We may continue to add more and more magnets. When we do this the magnetic energy combines to form still only one magnet. At the direct physical center of all the magnets, regardless of how many may be used, we find the Bloch Wall has come into existence. The Bloch Wall, as we have shown earlier, is the mid or center of each and every singular magnet. It is the point where the magnets energies alter their spin, and each pole then spins in an opposite direction as to its electron movement.

The next interesting fact is that should you in any way break a magnet in two parts, each part then becomes a two pole magnet, regardless of the size or shape of the two pieces. Each part forms its own two poles—the north and the south pole.

Can a magnet's two potential energy forces provide us with clues to its relationship with the force field we know as gravity?

Is a magnet's energy a form of atomic energy?

If chemicals and other materials can be magnetically polarized, will their weight change?

The answers to these questions now appear to be a very definite yes.

In answering these questions our explanations are based on actual laboratory experiments that resulted in additional new information relating to how magnetism affects all matter, fluids, airs and/or gases.

For our introduction to our answer we present a drawing on this page. This drawing was prepared to show that the magnet's electron spins will not combine or merge. Therefore, they are not the same in nature nor are they homogeneous in energy form.

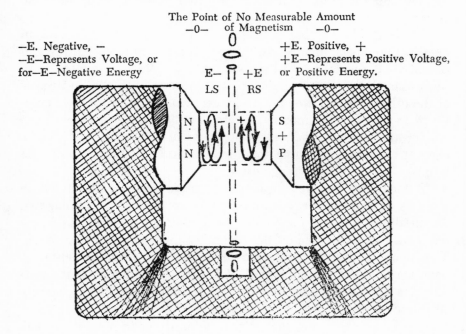

The Point of No Measurable Amount
—0— of Magnetism —0—

—E. Negative, —
—E—Represents Voltage, or
for—E—Negative Energy

+E. Positive, +
+E—Represents Positive Voltage, or Positive Energy.

E— +E

LS || RS

N
–
N

S
+
P

Note that in the drawing the north pole's energy spin is to the left, while the south pole's energy spin is to the right. These two energies cannot combine to form a unit of energy as we are led to believe in modern text materials. We have expressed voltage above as a possible first effect energy transfer. We have discussed earlier that the cables, or lines of force, travel in two directions not one. Voltage is shown by the letter E and the letter I for current. There has yet to be presented a suitable word for this dual force that acts as voltage and current does in electrical systems. The use of the term current, as expressed by I, is not the proper word, yet it will serve to identify the differences in part.

Continuing in our introductory material to the questions asked, study the drawing that is shown at the bottom of the page as the reason why the two poles of the magnet, their electron electrical energy spinning in opposite directions, fails to unify where they meet in the center of any magnet. That drawing indicates that at the center of each magnet there is a 180 degree phase change, a directional spin reversal. It is then at this point where the point of no measurable amount of magnetism can be recorded or measured. This point has been named the Bloch Wall.

However, should we study a drawing that has been published in a number of highly technical journals and books that attempt to explain this phenomenon of such a rapid, quick, unification of both reversal spin energies, we can see that while this is possible, it lacks the inversion of one energy to the other in the manner that energies unite under other similar unification of transmitted propagations as is shown in the drawing on this page.

THE BROKEN "8"

In the study of the unification of opposing potentials, we find on examining the energy there is the presence of the figure "8". We see this by making a visual study of the effects of the magnet's

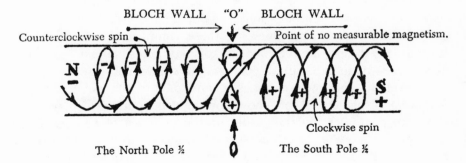

BLOCH WALL   "O"   BLOCH WALL

Counterclockwise spin   Point of no measurable magnetism.

N

S

Clockwise spin

The North Pole ½   O   The South Pole ½

energies on the screen of a large cathode ray oscilloscope. We will discuss this on page 97. At the left of the drawing are shown the left-hand spin and circulating lines of force that are present in the north pole half of a magnet, marked N. Next, see the point midway, O, at the center of these lines of force. This is the point of no measurable amount of magnetism, the Bloch Wall. Note that the figure 8 appears as a broken figure 8. The upper part of this broken 8 shows a negative magnetic force, while the bottom is shown to have a positive charge and also reverses the direction of the spin. From this point on to the right, we find we have a right directional spin, or the start of the south pole's positive energy.

Referring again to page 93, the drawing and the figure 8, we are aware that when energies are found in a natural state there is a division between each potential or charge. This division is equal. We believe the long studies of visual evidences we have made offer a totally new concept of how the energies of the magnet combine to afford a unification of the two energies, yet at the same time act to keep them separated. This we can see by the broken figure 8 in the drawing on page 93. The upper half of the figure 8 being negative in respect to the bottom half, which is positive. Next, the N pole energies act to affect a negative potential, and the S pole energies then affect a positive potential. Consider that this dividing line, as in the center of the figure "8" shown, is thinner than a human hair in width. When this is considered compressed, encompassing the entire spins of force, each spin each circle of energy, it taxes one's thoughts on the mathematical perfection, unity and compression we must envision to present this graphical picture to our mind.

Consider other electromagnetic visual projective investigations, such as in the study of modulated radio frequencies and the characteristics of wave propagations. Again, by the use of a suitable cathode ray oscilloscope, injecting these complex signals into the cathode ray scope, and the visual display on a screen, we can make some valuable visual comparison studies on the complex wave systems and their displays. Coupled with the possibility of magnetic displacement of known wave propagation display, we can gain a greater understanding of the energies of the magnet, their form, shape, energies and spin direction, as well as the overall projection of these magnetic energies from the magnet poles.

CATHODE RAY TUBE

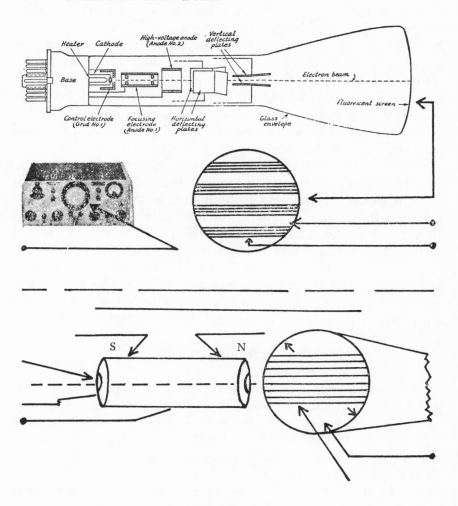

The cathode ray oscilloscope gives a visual representation of signals at both audio and radio frequencies and can therefore be used for many types of measurements that are not possible with instruments of the types discussed earlier. In radio work, one of the principal uses of the scope is for displaying an amplitude-modulated signal so a phone transmitter can be adjusted for proper modulation and continuously monitored to keep the modulation percentage within proper limits. For this purpose a very simple circuit will suffice.

The versatility of the scope can be greatly increased by adding amplifiers and linear deflection circuits, but the design and adjustment of such circuits tends to be complicated if optimum performance is to be secured. Special components are generally required. Oscilloscope kits for home assembly are available from a number of suppliers, and since their cost compares very favorably with that of a home-built instrument of comparable design, they are recommended for serious consideration by those who have need for or are interested in the wide range of measurements that is possible with a fully equipped scope.

The heart of the oscilloscope is the cathode ray tube, a vacuum tube in which the electrons emitted from a hot cathode are first accelerated to give them considerable velocity, then formed into a beam, and finally allowed to strike a special translucent screen which fluoresces, or gives off light at the point where the beam strikes. A beam of moving electrons can be moved laterally, or deflected, by electric or magnetic fields, and since its weight and inertia are negligibly small, it can be made to follow instantly the variations in periodically changing fields at both audio and radio frequencies.

The electrode arrangement that forms the electrons into a beam is called the electron gun.

To allow a certain amount of technical understanding of what a cathode ray visual viewer system is, see the drawing, which provides a very primary basic design of this tube. The cathode ray tube was the first invention for display of electromagnetic energy and is the forerunner of the present-day large television tubes. The short description of this tube will allow a fundamental understanding of this visual means of examining electromagnetic energies that otherwise could not be seen by the human eye.

As shown in the drawing, to the left end of the internal parts within the tube proper, the section marked "Control Electrode (Grid No. 1)," is where the signals or their effects are introduced into the tube for visual display on the fluorescent screen located at the far right end of the tube. This is the viewing screen.

It is a fact that externally applied magnetic fields will affect the cathode ray beam generated by the electron gun as it strikes the screen. This is done by placing one end of a cylinder magnet or straight bar magnet directly in contact and against the outer front surface of this tube. However, before doing this, supply the internal grid control with a horizontal bar test pattern from a television bar signal generator. We will see the horizontal and vertical bars appear on the screen. Then, bringing up the end of the magnet to the screen where the bars, both vertical and horizontal, are displaced they form an outline of the magnetic energy that is now passing through the glass screen and deflecting the pattern displayed on the screen. While this allows us to see with our eyes the spin of each pole, it also presents many new theories and concepts about the core of the magnet and its energies that are focused to extreme fine lines of force we have designated cables of force.

The discovery of the figure "8" was made by refinements to the above simple method in seeing the energy effects on the tube's screen. This geometric display prompted us to go deeper into the fact that at the Bloch Wall there existed a balance as well as the change in the unity of the pole spin direction. There is a neutral state at the Bloch Wall, as can be seen by the presence of two half segments of two energies. This then cancels out either energy, forming the Bloch Wall effect. During the instant cancellation there is also the point where one energy ends and the other begins.

In our developments with still further advanced systems that use the cathode ray electron gun principle, we have discovered that each end of a straight bar or cylinder magnet produces a cone shape of emitted energies. While very similar to an ice cream cone, its energies are not quite so pointed at the small end. The small end emits from the magnet's pole to grow in size, yet keeps a geometric perfect circle as it extends outward. We have measured this cone effect for many feet from a conventional magnet. It shows no point of return to the magnet. The straight line projection from the core center of a magnet is again a factor of great importance since this is also not traceable in a return to the magnet. This is a different cone effect than we first mentioned. Further, in discussing laws of the magnet, should we take a bar or cylinder magnet—long or medium long—that has been made with a hole running through its entire length, we would find in using the cathode ray contact to the screen, methods previously described, a compression of electron energy within the core of this magnet that would surpass any artist's dream of perfection. There also is a point where energy is focused down to an almost microscopically small dimension of constant energy. This is still to be explored further. The technical name for the figure "8" is the Lissajous electronic display projection figure.

Geometric and mathematical laws show us that a 360 degree phase shift is possible with open-ended circles, as we have discovered with the figure "8", in complete separation of the two energies. This may also be computed to give us a fractional minus error, yet is very close in exact calculations.

There should not be any question in the reader's mind by now that both poles are different in potential and spin direction, which identifies the energy potential differences and the effects each can have on living biological systems. All of these experiments, outlines and development discoveries are reproducible by anyone, at any time, at any location, if these persons are so equipped to carry out the experiments properly.

CHANGES IN GRAVITATIONAL WEIGHT

On page 99 we show a number of drawings to better outline and describe the effects of gravity and magnetism on the physical weight of chemicals, fluids and solutions.

At the top of page 99 we show a large commercial electromagnet connected to its power supply which converts A.C. voltage and current to D.C. voltage and current. This supplies the large electromagnet with D.C. current that makes the electromagnet a D.C. electromagnet. The effects of an electromagnet are not the same in many respects as the energies of a solid state metal or composition magnet. This we have found in other experiments which we may describe in detail in a further publication. Another D.C. electromagnet with power supply in table console is shown on page 99.

We now show a test tube three-fourths full of water or chemical solution or any solution of a fluid nature. It need not be a fluid but we chose a fluid for discussion.

The exposure of and treatment to fluids, liquids and proteins by this method has provided us with a great deal of vital research information that is far-reaching.

You will note on the study of the position of the test tube that the Bloch Wall again appears and we have two forms of energies,

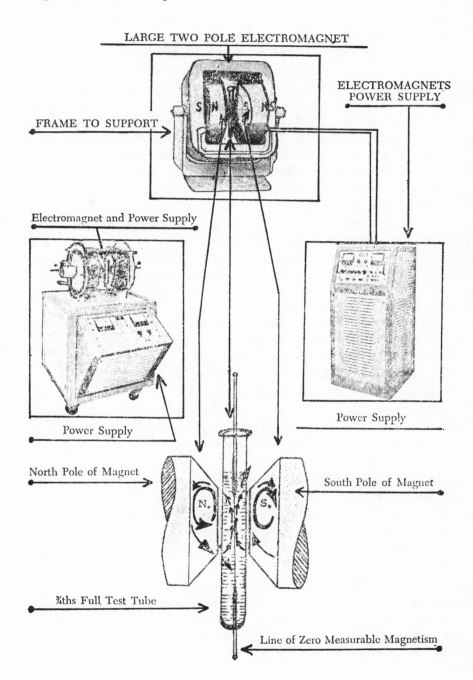

LARGE TWO POLE ELECTROMAGNET

ELECTROMAGNETS
POWER SUPPLY

FRAME TO SUPPORT

S N N S

Electromagnet and Power Supply

Power Supply

Power Supply

North Pole of Magnet

N. S.

South Pole of Magnet

¾ths Full Test Tube

Line of Zero Measurable Magnetism

each energy then treating or exposing the contents of the test tube with that particular potential of energy.

Next, you will note that the test tube is positioned vertical to the poles. As the pole piece electron spins affect a vortex of energy the force of this energy, the electrons spinning opposite each other, reacts to provide a pressure and a relieving of pressures in the vertical plane of effects on and to the contents of the test tube.

This acts to alter the physical weight when sufficient power of magnetic energy is applied to the sample between the two poles. This change in weight is in fact a change in the gravity. Many chemicals, fluids, proteins, not only alter and change their physical weight, their gravitational weight, but after a short time within the gap of the poles, act to alter the availability of the enzyme percentages, not in amount, but in digestive qualities. There is an increase in digestive results. Too short or too long an exposure destroys the effect. This may well account for the reason many of these experiments in the past have failed to be reproducible or failed in this respect from the beginning.

The S pole effects in part are to exert a pressure, while the N pole effects on the test tube contents are to draw away. Should we use a test tube of a size such as 8mm we would then find that at the exact internal center of the test tube a highly focused subminiature point of that pole's energies existing. There are many little-known effects that exist within the vortex of each pole. We have discussed only a few. When each effect is removed from the other, they act as forces that may be separated and used for a direct reason for desired results.

There are many chemicals that show a remarkable weight change when placed between the poles of the magnet. This is due in part to their active contents of ferric iron and other magnetic acceptable materials. However, inert fluids, chemicals and other materials also show this gravitational change in the overall physical weight results.

As an example, the NASA space probes provided us with similar reactions. When the space capsule left the earth's atmosphere, the sun's ray striking the sides of the capsule acted to supply a force, a pressure, that altered the course through space on the computed directional path of the space capsule. It was then necessary to correct for this photo-electromagnetic pressure effect.

Our findings show that each pole, therefore, applies to whatever it is facing a form of pressure, a spin pressure, that affects the gravity weight of the sample.

FUNNEL EFFECT OF MAGNETIC ENERGIES

In the most elementary application of primary physics, one might then compare the field of a pole of a magnet as an inversion of a tornado, differing only in the fact that in the development of a tornado, the narrow end almost touches the earth in its downward movements and the larger vortex is above the earth. The magnet's vortex starts small at the end of the pole and enlarges as it travels through space. The diameter of the vortex is equal to the distance from its pole end to the widest part of its extended vortex.

Within the wide diameter vortex of a magnet's pole is a similar attracting vortex spin. The magnet's spinning tornado-like vortex contains two expressions of power and energy. The electron spin acts as a force of spin direction, and the inner or reverse vortex located within the center of the electron fields acts to form an uptake of energy or a downward thrust of energy. The reversal of the spin effect results in the reversal of the inner vortex reaction. Like the tornado, the magnetic fields attract any atomic molecular substances that are within the acceptable range of elements that would be attracted. This differs from the physical atmosphere action when we refer the overall reaction to the magnet's vortex energy to the similar form reaction of the tornado.

One elementary question that is often asked by students is "Why does a magnet attract and draw to its pole or poles such things as a straight metal pin?" This in itself may not seem on the surface to be a very important question, yet it is a forthright question and would require a modern physicist quite some time to give the correct answer. We repeat—a correct answer. The fact that the atoms within a pin or similar material matter are in part set into motion is not the answer for the attraction; here we describe the magnetic pull to ferric or other magnetic materials by either or both poles.

The approaches we have used, and in part explained on page

97, show us that the fields of either pole have within or without their formulations of vortex energies the indraft action effect that is close on inspection to the vortex of a tornado. In the case of a magnet's poles, the fact that there are two directional spin energy vortexes, we believe, opens new doors to a greater under-standing of the atomic resultant molecular attraction of these compositive molecule arrangements in such attractable substances or matter. In simple language, it is the "funnel effect" of the two magnetic energies within the combined magnetic field that attracts materials having an atomic molecular attractive formulation, such as a simple pin.

A NEW WORLD OF SCIENCE

We could have filled these pages with mathematical formulas based on the textbook materials presented as the last word on the laws and principles of magnetism. This would have resulted in a book very valuable to a few, but the greater part of the scientific community and students would have been lost in a complex mathematical and geometric presentation. There have been more books written on magnetism than possibly any other subject in existence today. None offers the desired answers to the natural laws that are offered by actual physical and biological studies of the effects of magnetism on matter or on the living system itself.

We have found that what can be experienced to aid mankind in the biological effects of applied voltages and currents can be affected in a like manner with the properly applied magnetic pole energies of a magnet or magnets. In bone regrowth, fracture strengthening and knitting, by the inserting of needles into the bone, above and below the break and applying a low voltage D.C. and currents, the break is repaired in less time and is stronger than natural healing could have accomplished in many cases. Then by the proper application of the poles of a magnet or magnets above and below these fractures, a far better healing reaction can take place and quicker results. The control of pain and an overall faster better experience of desired results may be seen. Here, as we have stated before, the science of "Biomagnetics" needs no internal probing into the living system as does the UNNATURAL science of Acupuncture. Bone regrowth is also possible by the proper application of magnetic fields; this is opposed to the administering of needles attached to batteries to afford the same or, in fact, less desired reactions.

The science of biomagnetics offers modern science a totally new tool to advance all sciences today, unlike the science professor who told his class there is nothing left to invent so be satisfied with what we have to work with today. Research means to search and research those knowns against unknowns, and this science of biomagnetics is the very basis of all existing sciences known to man today. Its very fundamentals are based on the laws of the atom, and the atom is in itself a small magnet. The atom is the building block of molecules, and the molecules in turn are the building blocks of all matter—airs, gases, fluids, solids or intermediate materials. When you enter into the research and investigation of biomagnetics or magneto magnetic energies you are entering a new world of science, one that affords a greater understanding of all natural laws governing energy, its applications in and to all fields of science.

Should you be presently engaged in magnetic research, we believe that the new developments and advancements to the knowledge of magnetics contained in this book will assist you in your continued efforts, regardless of your interests in the scientific applications you direct these findings.

ATOMIC ENERGY AND MAGNETISM

We have discussed many aspects of magnetism and the atom and would now like to offer in evidence these supporting facts. First, when we apply a strong electrical D.C. voltage and current to a coil of wire and insert a piece of ferric or magnetically acceptable material into the core of this coil, we have polarized the atoms within the steel or magnetic material to align all in one direction. According to the textbooks we have made a magnet.

However, if by doing this we have in fact aligned all the atoms within the shells of material that are subjected to this polarization of the atoms within the material, have we not also by doing this developed a form of atomic energy? It would appear we have. Second, it is not necessary to form atomic energy at the high frequencies in the angstrom spectrum that we normally consider and think of as the high to superhigh number of angstroms, as past X rays and alpha, beta and gamma radiations. Atomic energy need not be the kind or nature we generally think of as having to be in these extremely high frequency spectrums. Atomic energy, as it is defined, is a source of energy that is the result of obtaining radiations from atoms in an excited state. Magnetism—the magnet's two poles, each has a frequency. We have discovered this and have been researching and studying the spectrum that these energies fall into for further clues of the effects of these frequencies in relation to the length, form and size of a magnet. While there is still much to be learned from these initial discoveries, we feel that we are nearer to the laws that govern magnetic and magneto energies, basically a form of atomic energy, yet at this time not considered in this regard. However, there is no doubt now that magnetic magnet magneto energies are closely related and are similar in many ways to the higher forms of atomic energy.

We believe we have established the relationship between magnetism, electricity, gravitation and atomic energy structures, thereby demonstrating a basis for the unification of these energies.

MAN BOMBARDED BY ELEMENTS

In the chapters we have presented so far, we have discussed many effects of magnetism on the living biological systems, and this encompasses mankind. All living systems are affected by the earth's magnetic fields. Earlier we mentioned that man is bombarded by the elements. The importance of this matter deserves a separate subtitle and further discussion.

All life forms are affected by the outer radiations of energy of the planets in space. Those nearer to us, like the sun and the moon, exert great pressures and electromagnetic effects on these living life forms, as do the resultant gravitational pulls and pressures that these outer space forms of energy apply to earth and all material matter. This includes waters, airs, gases, solids and semisolids. Nothing escapes these unseen forces. Man's environmental attitudes are affected as well as his or her physical feelings of welfare, even to his or her general health patterns.

As an example, in the Soviet Union they have built special

shielded rooms that resist the effects to a high degree of outer magnetic influences. They have discovered that, as an example, persons suffering from nervous and other complaints, including forms of heart disease and those suffering from mental ailments, all react adversely to the effects of sunspot storms and changes in the gravitational effects of the moon. The Soviet scientists then change the medications because of the effects these outer electromagnetic and gravitational energies have on the sick and the weak. We should look into this. It is an important matter and one that has been overlooked in many countries far too long.

To better show the ionosphere man lives in and how by its changes these changes affect the health and well-being of all living systems, including plants and even microscopic organisms, we present a number of simple drawings on page 106.

Drawing A shows by the shaded areas the ionosphere. This is one of a number of such layers that surround the earth at varied distances from the surface of the earth. They act in part to filter out harmful rays from space.

As they then fulfill this duty they also act to keep certain energies within this protective shield as well as down on the surface of the earth proper.

These natural ionized layers vary in height, and this varying of height then acts to influence radio waves and TV signals that on leaving the earth by the transmitting of these electromagnetic energies we call radio frequency waves. From the station antennas they go upward, and when they reach the first ion barrier they are reflected back to earth. This is, in part, how radio waves, radio signals and programs are able to reach so many locations within the scope of the power of the station and the frequencies used. This is broadcasting electromagnetic wave propagation to the listeners and viewers who have suitable receiving sets or equipment.

In the science of probing space, the transmitters are designed to penetrate these ion shields, the ionosphere and its many layers, to go out into space for this type of radionic exploring. These very high frequencies are also returned through the ionosphere back to large radio telescopes for detection and computerized data results.

Refer to drawing A on page 106 (top); the shaded area entitled

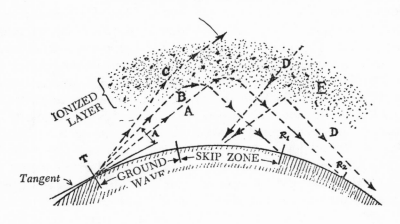

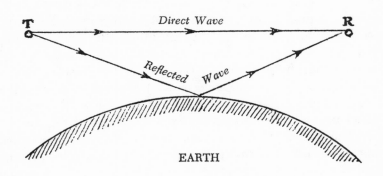

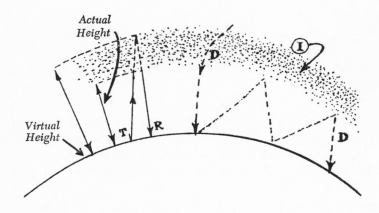

"ionized layer" presents a graphic example of the many such ion shields that surround the earth. Dotted line C is to show that at extremely high frequencies, space-type exploring transmitters may penetrate the ionosphere. Dotted line A shows how the conventional medium frequency range broadcasting stations signals are reflected and bounced back to earth. The points of the reflection and return path to earth are shown by the dotted lines. In between these bouncing and reflecting signal paths are the zones of reception, in part. This is the wave phenomenon propagation that explains the radio signal's ability to reach so many viewers or listeners.

Further, we note that in drawing A we show the "skip zone." This lies between the reflected waves. Reception can be very poor and signals difficult to receive that are in this zone or skip area. Yet it has its advantages also, as by the ability to skip, the range, for example, of a medium short wave (medium high frequency) radio or TV station can be greatly extended. And stations that normally cannot be received can then be received when the ionosphere layers above the earth rises or lowers. These ion shields act in part like electromagnetic mirrors, reflecting back to earth signals that otherwise would go into deep space and perhaps never return to earth.

Next refer to drawing dotted line D in the same drawing A. This represents ions energies. They can and do penetrate the ionosphere shields that protect mankind and all living systems from the harmful energies of the sun. They deflect energies from the moon and other planets in outer space. Some energies do get through to the earth as shown by the dotted line D.

Drawing B is presented only to show the geometric wave propagations. Drawing C on page 106 is a simple extension of drawing A to show the reflective abilities of the ion layer effects.

We can now see that the ionosphere then acts to protect all living systems from the deadly rays and energies that would destroy life as we know it if the ionosphere was not acting as a protective shield around the earth.

Within the earth's ionosphere, the lower atmospheric ionic atomosphere, is the density of positive energies, such as the massive positive charges of positive ions that are present in clouds. And when storms result, when the clouds become supercharged, the positive ions act to make their presence known to man and living systems by the increased density of positive ions in the atmosphere. Man's health and that of animals and also plants are governed by these atmospheric charges. Mental depressions are one result of too much positive ion concentration. While in a negatively charged environment man's welfare, outlook on life, mental abilities are upgraded. The negatively charged ions a person finds so pleasant then act on the living system when diseases, complaints and weaknesses are affecting his or her health. A better recovery of many illnesses can be obtained in a negatively charged environment. The positively charged environment acts to slow recovery, affects the lowering of mental activities, making it more difficult to think and function. These two forms of atmospheric environmental charges or energies reflect the same effects as the north and south poles of magnetic energies. We have proven this time and time again in animal and rodent research, including large animals that have similar blood types to man, the same organs and life functions. The environment man and all living systems find themselves in from day to day, hour to hour, is in constant change. Man is in effect a "Biomagnetic Animal in every respect."

Chapter Twelve

THE HUMAN BIOMAGNETIC AURA

Man, animals, plants or bacteria all have a separate and different biomagnetic aura. This was taught long ago by the ancients and is today subject to scientific proof. Each living thing has the ability to transmit from its body certain electromagnetic energies. The ancients believed that persons who were sensitive to these colormetric energies could see a flow emitting from the body. By such means they could then identify illnesses from the change in colors presented by locating diseased or infected nonoperative organs and parts of the body. This art is still practiced all over the world.

The scientific community has generally refused to investigate this phenomenon, placing it in the background as more of a belief in the supernatural rather than having scientific supportable fact. However, this may now change as we have scientifically proven that the living system, regardless of its nature, emits an electromagnetic energy.

In determining these energies we have used modern electronic detection and recording methods to measure and record these emissions from the body of the living system. By the research into electromagnetic energies and their effects to the living system, we have also researched those electromagnetic energies that living systems emit. This should change the old concepts that this possibility exceeded the power and laws of nature as the words *superstition* or *supernatural* imply.

Our work that follows was placed on display at the Westfield Conference in London a few months ago and was received with great interest by the scientific research community, including medical and applied science researchers. We shall now present our proof that the human and/or living system can and does emit electromagnetic energies of many different frequencies.

109

THE LAYING ON OF HANDS

Many scientific men and women in research today have discredited the possibilities of any scientific proof, overall effect or reaction of the ancient and recorded healing approach, "The Laying on of Hands."

It is our opinion, based on scientific findings, that this science will prove to be a valuable contribution to mankind.

Within the text materials to follow we shall show the recorded electromagnetic energies on the area of the body of man. These recordings also apply to the body of woman. We will also show the potential power and force that is present in the palms of both hands and fingers. The human body is a vast complex of energies and frequencies. We can compare the palms of the hands in their application that affords an exchange of natural body energies from one person to another. We will show by charts in this chapter the different potential force in each hand should that hand rest on the body of a person suffering from a pain, fever or complaint. There would be a proper reaction and a transfer of energies. Should we wish to become more technical we would say there is an inter-exchange, a modulation, of both body energies.

It has been shown that when a living system is weakened by the effects of sickness that the body is weak also in the average electrical energies at the location of the difficulty that the person is then experiencing. Should a healthy person apply his correct palm to that ill person there is an exchange of normal energy to the lowered energy, and a feeling of well-being is experienced by the person who is ill. We do not say cure. We do, however, make this positive statement as to recordable facts. There is a feeling, a sense of comfort, of well-being, or of relief.

The human study of a mother's care and attention of her baby or the child presents to us that inborn inner instinct of caring for and offering love, devotion, strength and providing the baby or young child with every possible help with or without the mother's knowledge of the "Laying on of Hands."

It has been of great help in our research to study the care of the baby or child by the mothers. One of the observations made was that when the child is sick, ill, or suffering from an ailment, such as having a slight fever, the mother will rest the back of the right hand on the child's forehead. Now, what is happening, and probably unknown to the mother, is that she is applying the negative energy, the negative microvoltage that exists on the back of her right hand, to the fevered forehead. In our years of research into the human behavior of man and animals, we note this happening takes place and the results of this laying on of hands is one that offers the baby or child a high degree of comfort. This has been observed hundreds of times. It has resulted in the same effect, a calming, soothing reaction to the subject.

Many mothers will apply the palm of their left hand to the subject's forehead. There is also at that point a full negative electrical, or electromagnetic, energy present as the palm of the left hand is negative opposed to the palm of the right hand that contains a positive energy potential.

In the presentation of the charts of the outlines of the human body and its extended members we shall present in this chapter, the recordable voltages also encompass electromagnetic energies and a modulation of varied changing frequencies.

Further studies have shown that on recording the human voltages the back of the head has an external measurable positive voltage. The mother has shown us that when the baby is weak she places her right palm under the back of the baby's head. This is applying a positive electromagnetic energy to the upper spine nerve column and the back of the baby's head which has a natural positive energy potential or charge.

Here we see that while the mother is not an advanced student of science she has inborn natural talents and knowledge that rest in the subconscious mind. These inner and all natural instinctive reactions will provide the exact and proper energies that flow between the mother to the child as the need may be.

Here, for a point of reference, we note that the palm of the right hand offers a positive energy opposed to the palm of the left hand that offers a negative energy. Reversing the energy, we see that the back of the right hand is negative opposed to the back of the left hand which affords a positive energy. It is of the utmost importance in furthering the understanding of this natural and applied energy that we be fully aware of the subject's own natural energies. For example, when a bone is broken, wherever it may be located, the body's own natural electrical system then affords to that exact point of the break a much higher than normal negative electrical energy. This returns to a normal energy level only after the bone has reknit and healed. Thus the body supplies the correct energy as to positive or negative according to its demand to offer help and assistance to its own part or segment that is experiencing trouble.

In our research as to the effects of magnetism and its two separate pole energies to the living system to afford help and assistance, we discussed the north pole of a magnet supplying a negative electromagnetic form of energy. This is opposed to the south pole of a magnet that offers and supplies a positive form of electromagnetic energy. Again we see that a magnet's energies are an all-natural energy form similar to that existing in and from the human system.

During the association with Dr. Yerkes of the Yerkes Experimental Research Laboratories, then a division of Yale University, your senior author noticed the same applications of the hands of the monkeys, apes, and species of animals in that category, were applied to their young when sick, weak or ill. This points to the fact that even these species of animals react similar to the human mother. Each has a natural built-in instinct of what to do to help and care for their young.

THE HUMAN AURA

Here we examine the existence of an energy form that surrounds the human body and exteriors of all living systems.

Research studies released some years ago by the Eastman Kodak Company are helpful in our presentation. These studies dealt with the dark red light of special lamps, infrared frequencies, as applied to the naked body of man.

When the body's outer exterior was radiated with this almost invisible frequency photographs were taken in a few minutes with infrared film in a camera. In the photographs, appearing as black and gray lines, were the arteries at the surface of the body and large exterior and interior veins. The further distance from the heart and lungs that the blood circulated more oxides were formed in the blood. This oxide increase was due by the use of the oxygen taken from the blood by the body's functions.

Where the body had a poor circulation, the pictures presented that part as dark gray or black in color. This was due to the lowering of the body's temperature at that part of the body. Due to the lack of heat at that part the infrared film picked up this difference in temperature on the body's surface and it was recorded on the film as dark to black. This weakness in the body system, shown and analyzed by infrared photography, is evidenced by the same lack of color, black, sensed by those who see the Human Aura, as a weakness in the human system.

Certain people have far more sight sensitivities than other people. The authors do not pretend to understand all of the factors that assist the extraordinary ability to see the Human Aura. The same is true about the ability to practice the "Laying on of Hands" effectively. However, these abilities do exist and it is a fact supported by scientific experiments and findings.

To further support there is surrounding all living systems forms of energies that can be recorded as fact and not as theory or superstition, we submit from our laboratory research the following findings. We have measured from a fraction of an inch to greater distances from the surfaces of the body energies that are in a sense a radiation of electromagnetic energies. This statement is made in this broad manner as these energies are composed of many different forms, frequencies, microvoltages and currents. This does not change the fact that they are present, projected from the surface of the body at all points. Following the curvature of the physical body proper, again we see the similar form which those who state they can see this energy form describe as emitting from the body. A number of these gifted persons tell us it is possible to see these energies surrounding the human body, and even see the energies surrounding any form of living system such as animals and plants. This they advise can only be done in subdued light. However, the outline of the energy they describe follows our basic scientific research measurements and resultant findings that can be reproduced at any time anywhere under those certain light subdued conditions and environmental surroundings by the proper trained person.

COLORS AND THEIR EFFECTS

Colors are composed of frequencies of energy. Each color as we see these energies with our eyes differs only by the number of cycles or vibrations per second each color has or contains.

The Color Red, in actual agriculture research experiments, acts to promote health, life and energy. This was confirmed in experiments where small growing potted plants were placed under these color lights. Red gave life, energy, and a higher degree of growth and development. Some of these similar experiments were conducted at the universities in California a short time ago.

The Color Green is the life-controlling color, the reducer, the arrester. Plants placed under this color acted to present positive scientific laboratory resultant facts by controlled experiments.

The other colors are moderators of the two colors, Red and Green. Comparing the colors Red and Green to the forms of energies as presented by the two separate and different poles of a magnet, we find that Red is like the S pole energies and Green is like the N pole energies of the magnet. Positive energy in whatever form acts to promote life or it can overpromote life to a point of depression. Green, as the N pole of the magnet, acts to arrest life in the control of life forms. Here again we see that all life on this earth of ours responds to the two forms of dissimilar energy.

In the latter pages of this book we will show where the positive and negative energies of man are located. We note again that we can no longer consider or abide with the old, yet current thinking, that the ancient sciences hold nothing more than superstitions. Far from this line of thinking is the need for further and extensive additional investigations into the ancient sciences.

MAN'S BIOMAGNETIC MIND

The brain and the mind of man differ in that the brain is a physical part of the body, this opposed to the mind which is the active working, thinking, feeling, results of the brain's functions.

Considering this it is in part the mind that acts as the computer and the brain the mechanics of that mind or computer.

Both the mind and the brain are electromagnetic in nature. They are electrochemically operated. The brain is like a small radio transmitter, and the mind is the voice that is modulated on the carrier of the transmitter. Therefore, when we examine both we can see the results, a thought-controlled small biochemical activated energy supplied transmitter of conscious thought patterns with the back-up transmission of the subconscious mind.

We have measured the brain's output of electromagnetic waves of energy. The brain has a number of very low frequencies that are generated within the brain. These low frequencies act to carry the mind's thoughts and reactions to the varied parts of the body of man and animal. The same is true in lower creatures. We find the frequency number 7 (7 cycles per second) as a good basic fundamental frequency the brain affords the body that can be measured and recorded on the surface of the head of man.

Man's mind can act as a receiver as well as a transmitter of thought information. The study of thought transfer is under serious research in a number of countries today. This was prompted in part by the space age research efforts. In Russia serious research has been going on for some time concerning the possibility of men on earth and in space being able to transmit key signals from one to the other. Both Russian and American space research has now found this is possible. Therefore, the brain is, in fact, a transmitter and also a receiver.

The subconscious mind has long been a mystery to man and to scientists alike. The study of animals and their natural reactions to dangers, the sense of dangers present yet quite unknown to man from the lowly mouse to the large species, the results of the subconscious mind's ability to override the conscious mind and so to speak ring an alarm bell within the conscious mind, is very real in every respect. We find that the subconscious is a higher frequency mind, one that is not burdened by the work load that is placed on the mind by the thinking mechanics of man's brain or the animal's brain. This higher mind, this higher analyzing mind, can sense and feel environmental happenings and be free to analyze matters, conditions, not only those happening at that time but also to foresee in time what might happen as a result of the actions of the personality and reasoning of that person. We base these projections on what we have observed over many years in the study of lower animals to the super reasoning anthropological species. Man is somewhat less developed in the natural effective sensitivities the animals show as man's environmental surroundings have taken away the inborn development of man's "Early Warning Radar" scanning network, so to speak. The animals, however, like man, are not always in the state of alert. Yet there are certain animals that operate a 24-hour-a-day early warning system. The fewer defenses the animal has against its environmental dangers, generally the greater its ability to sense dangers.

As is the conscious mind, the subconscious mind mind is just as important to live and survive. To have only a conscious mind would remove any sense of dangers or detection. There could be no instinct for the future and probably no inspiration. Certainly, the subconscious mind plays an all-important part during the life span of animals and mankind.

PARAPSYCHOLOGICAL-BIOMAGNETIC INVESTIGATIONS

What part does magnetism, electromagnetics, play in the field of ESP (extrasensory perception) or the science of parapsychology? If we examined the term extrasensory perception we will or should find it incorrect as to its meaning, as all living systems have a sensory perception. This we have discussed in the last few pages. The "extra" term then would properly mean the development of the senses or sensory perceptions that are an inborn quality the living system possesses.

Can external magnetic energies when applied to the animal or man act to encourage or depress the mental sensitivities of the mind and the functions of the brain proper?

Earlier in our book we have shown the effects of the two pole energies of a magnet on the embryo, the newborn creature, and during its stages of physical development. Indeed, our findings indicate that externally applied energies can and will affect the development of the mental processes of the mind and the brain of all living systems.

It is with deep regret that we read reports from outstanding scientific investigators in their books, papers and reports that deal with this subject. They discuss the only effect they see from the placement of a magnet against the head of man or animal is a sharp flash of light, seen at the edge of the eye on placing and removing of a magnet's energies. This report of effects is also to be found in current space agency research reports. This can be a very dangerous as well as incorrect conclusion.

In research experiments with small and advanced animals and man, in the case of willing subjects, we have found that the magnet's NORTH POLE ONLY, when applied to the brain, can and will upgrade the senses of perception.

Perception here is meant as the overall normal inborn senses of animals and man alike. We find that the north pole, which is the negative energy as its effects to all biological systems, can and does possess the ability to pass through the head, the skull of man and animal, and act on the inner constants of the brain. This acts to cause a stimulus to the hemisphere of the brain, the reasoning mind itself. In our studies of this interaction we have discovered that when this form of energy passes unhindered into the innermost part of the brain it acts to arrest certain pressures on the brain by a reduction of inner and outer electromagnetic reactions to the brain. This relieves biological biochemical pressures as such from the thinking and reasoning mind. This then as a result has shown the ability of the man to think, reason more clearly and has aided in the learning and capacity of the mind and the storing of that learning in the mechanical segments of the brain. This statement may appear as being unsupported by other scientific findings. However, our investigations have been extensive in exploring the brain's and the mind's reaction to magnetic fields. There should be prompt serious investigations as to what can be done in this field in the future. Many governmental agencies of a number of countries, including the United States, have been for some time upgrading their research and investigations into ESP and parapsychology as a means of interspace communications and also into other phases of use in fields of mental telepathic data transfer.

When we consider that the mind, the brain as such, does send and receive intelligences by electromagnetic energy waves, we can see how important this science of magnetics is to man. Your authors predict that the future use of magnetic energy applications will result in a new and better understanding of "parapsychology" in science.

WARNING

There is within this science certain dangers that must be considered. In the use of S pole positive energies in brain research we have discovered that these energies can overstimulate the brain and the mind. Also, there may be hidden defects resting in the physical or mental aspects of the brain and mind, such as growths or types of abnormal behavior, which the S pole energies will encourage in unnatural development. The S pole energies in brain research should only be used by properly trained persons learned in this new science of magnetism.

At this time we also emphasize the dangers existing on improper use of either the S pole or N pole energies. Without the proper understanding and training necessary, the use of either energy on a living system can be dangerous and ultimately fatal. Students, physicists, medical men, scientists, should exercise caution in their work and the same warning here, and more so, is given to laymen who read this book.

Your authors have taken years deciding whether or not to release the findings in this book. The decision to release these findings was not an easy decision. Science can only progress by its development and application of new discoveries. Men may use or misuse science. The nature of man still resides in his behavior to his environment. For all the danger in this new science of magnetism your authors must look to all the good the science holds for humanity. Your authors live with their decision as humanity must live—or die—by its decisions.

MEASURING THE VOLTAGES OF THE LIVING SYSTEM AND THEIR LOCATIONS ON THE BODY

On page 119 we present an outline of the human body. For our purposes this outline and subsequent outlines in this chapter will also be a reference to the female body and to animal bodies. Our laboratory experiments and findings have shown that the

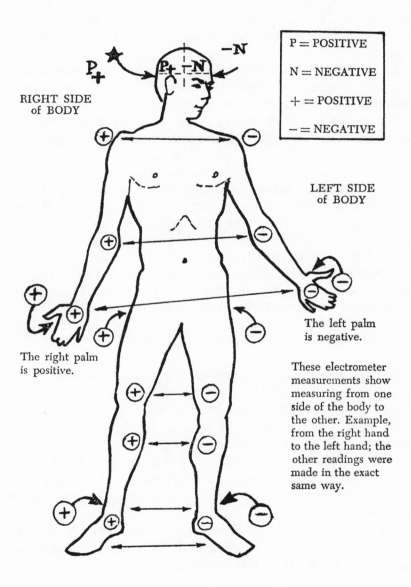

RIGHT SIDE
of BODY

LEFT SIDE
of BODY

P = POSITIVE

N = NEGATIVE

+ = POSITIVE

− = NEGATIVE

The left palm
is negative.

The right palm
is positive.

These electrometer
measurements show
measuring from one
side of the body to
the other. Example,
from the right hand
to the left hand; the
other readings were
made in the exact
same way.

surface voltages indicated give interesting research information
in determining where to place magnetic energies near the human
body. Other uses of these recorded voltages in research may be
apparent, though we confine our discussion here to our expressed
subject.

As we have stated earlier you do not just put a magnet to the
body and expect the desired results. If magnets are used for
desired results they need to be of special construction, shape and
size. The strength of the energies varies for particular results. As
important as the strength in gauss units is also the length of time
necessary for the specific application.

Studying the charts in this chapter the sign + or P means
we have a positive voltage at that location on the body. The sign
— or N means the negative voltage at that location on the body.
For example, we show the electrical potential in the palm of the
right hand as positive while the electrical potential in the palm
of the left hand is negative. We measured the voltage of each
by attaching two electrodes, one to each hand. The positive and
negative polarity is then registered on the dials of our sensitive
electrometer instrument developed for this research.

BIOMAGNETIC BIOLOGICAL ELECTRONICS

Referring to the chart presented, our findings indicate that the
extended members of the right side of the body are positive
electrically. This applies to the hands, arms, legs and feet.

In looking at the head, the front of the head is negative and
the back of the head is positive.

It is of importance that you notice that the right hand and arm,
although positive, in respect to the left hand and arm is negative,
carry still another charge at various locations in the bone structure
of the members.

This first chart shows lines crossing the body in a straight
direction with an arrow at each end, for example, from hand to
hand. This is to show the measurement of voltage was taken
from hand to hand, from the one to the other in relation to each.

In the second drawing of this chapter, on page 121, the right
hand, arm and side are indicated by the large letter R and the
left hand, arm and side by the large letter L. This drawing shows

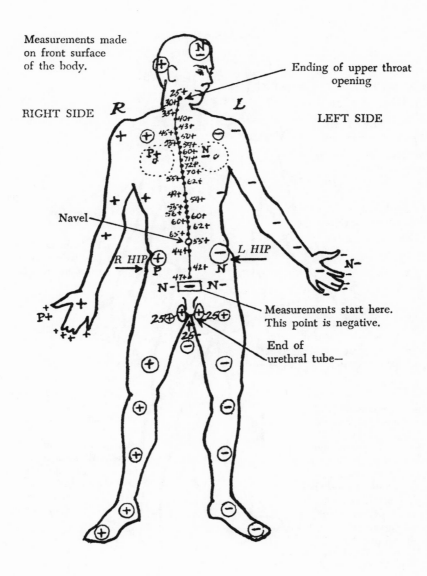

Measurements made
on front surface
of the body.

Ending of upper throat
opening

RIGHT SIDE   *R*   *L*   LEFT SIDE

Navel

R HIP

L HIP

Measurements start here.
This point is negative.

End of
urethral tube—

Back of the Body

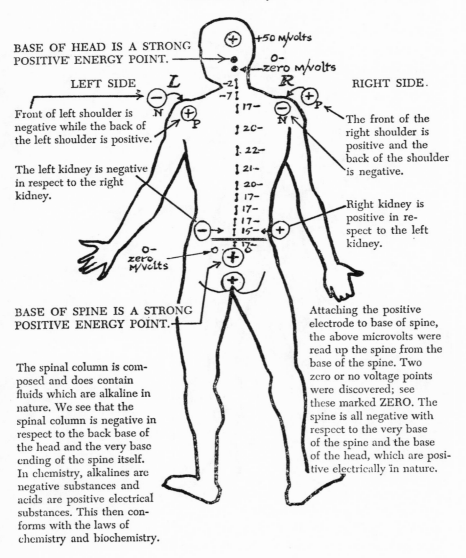

BASE OF HEAD IS A STRONG
POSITIVE ENERGY POINT. ——→

+50 M/volts

0-
—zero m/volts

LEFT SIDE *L* *R* RIGHT SIDE.

Front of left shoulder is
negative while the back of
the left shoulder is positive.

The left kidney is negative
in respect to the right
kidney.

-2
-7
17-
20-
22-
21-
20-
17-
17-
17-
15-
17-

The front of the
right shoulder is
positive and the
back of the shoulder
is negative.

Right kidney is
positive in re-
spect to the left
kidney.

0-
zero
M/volts

BASE OF SPINE IS A STRONG
POSITIVE ENERGY POINT. ◀

The spinal column is com-
posed and does contain
fluids which are alkaline in
nature. We see that the
spinal column is negative in
respect to the back base of
the head and the very base
ending of the spine itself.
In chemistry, alkalines are
negative substances and
acids are positive electrical
substances. This then con-
forms with the laws of
chemistry and biochemistry.

Attaching the positive
electrode to base of spine,
the above microvolts were
read up the spine from the
base of the spine. Two
zero or no voltage points
were discovered; see
these marked ZERO. The
spine is all negative with
respect to the very base
of the spine and the base
of the head, which are positive electrically in nature.

the average voltages taken at various locations of the body. In the male figure as shown measurements are given from the crotch of the body a little below the end of the urethral tube ending, and the readings are also shown for the male testicles. In the female body a negative reading is slightly above the urethral opening. On the chart as shown for the male the testicles are positive, the urethral tube is negative, thus the negative plate of the electrometer was placed at this lowest point.

As we go upward on the chart we see the various voltages that were recorded, the highest frontal voltage measured at the point near the heart, shown as 70 positive microvolts on the diagram.

These voltages can change from one person to another. This is very important when considering the many applications from these discoveries. What we show are average voltages from testing a number of subjects.

On the charts we see that the spine is negative and that the exact center of the front of the body is positive energy. The frontal crotch is negative with the back base of the spine, which is positive.

Looking at other relations we see that the frontal central length of the body is positive with respect to the frontal lower crotch. The forehead or front of the head is negative with respect to the back of the head, which is positive electrically. The readings are all direct contact surface voltages, checked time and time again for polarities at specific locations.

Of interest are two locations where no noticeable voltage readings were found. One is shown above the rectum near the base of the spine, and the other location is shown at the connecting link between the skull and the backbone proper, the beginning of the full-length spinal column.

These two zero points on the human body are the electrical equators of the body's electrical and electromagnetic divisions. At these points is where the polarity of the voltages change from one form to another form. For example, we have a positive voltage, then a lessening of the voltage to a no noticeable recorded voltage before a swing over to negative voltage. Notice the similarity to the straight bar magnet and the presentation of the broken "8" discussed earlier. This recording of the two areas with no noticeable voltage, polarity, or energy is also similar to the magnetic equator of the earth as explained earlier in this book. For example, the S pole magnetic energy of the earth goes into a Bloch Wall effect at the near center of the earth, where it changes to a negative or N pole energy that continues on to enter the N pole of the earth. Similarly, there is a two-way highway system with the N pole magnetic energy traveling to a magnetic equator, where it changes to a S pole energy and continues on the highway to the S pole.

Notice from the charts that two or more polarities can exist on one part of the body, such as the leg, hand and arm. The inside surface palm can be positive while the outer side of the hand is negative. Also, the fingertips' inside surfaces can be positive while the side that support the fingernails is negative. This system also applies to the arm, as both right and left arms and hands can be negative while the inside surface of hands and arms are positive.

The drawing on page 125 shows the location on the frontal human system where no noticeable voltage was registered. As stated earlier there are also two similar locations on the back view of the human system, one near the base of the spine and the other between the skull and backbone.

You will notice that these are separate voltage systems in each hand, arm, leg, even into the surface bone of the leg proper. This separate voltage system continues throughout the body system. For example, the more alkaline, such as the bones of the spinal column, a negative potential; while the more acid, such as the flesh, a positive potential.

We can generally summarize by saying the front of the body is negative and the back of the body is positive. These two natural and living electrical potentials, positive and negative, enter into magnetic equators at two locations on the front and back parts of the body, continuing out of these magnetic equators in their changed potential, a repeat process over and over throughout the human system. In this brief outline of the bioelectrical potentials found to exist in man there is evidence of human electrical and electromagnetic aura. The entire body of man is a field of continuing flowing electromagnetic energy, and the space emissions of this form of energy have been recorded.

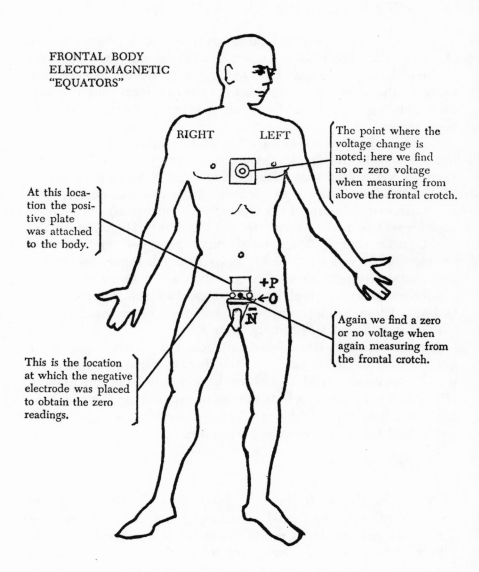

FRONTAL BODY
ELECTROMAGNETIC
"EQUATORS"

RIGHT   LEFT

The point where the
voltage change is
noted; here we find
no or zero voltage
when measuring from
above the frontal crotch.

At this loca-
tion the posi-
tive plate
was attached
to the body.

+P
←0
N̄

Again we find a zero
or no voltage when
again measuring from
the frontal crotch.

This is the location
at which the negative
electrode was placed
to obtain the zero
readings.

THE PRESENT AND FUTURE RESEARCH INTO
VERY HIGH GAUSS MAGNETIC ENERGIES (VHG)

While past and present research has shown little positive effects of VHG (Very High Gauss) energies to many biological systems, we should remember that the old concepts rule the science and its applications. We define old concepts to encompass the research understanding that both fields of a magnet are the same or homogeneous.

While it has been necessary for us to research the effects of the two pole effects using low to medium ranges of applied magnetic energies, we believe that with the basic understanding presented for the first time in this book concerning the two pole effects, with a greater understanding to overall effects, further significant discoveries will be made in the use of VHG (Very High Gauss).

Magnetic energy "Magneto Magnetic Energies" in the range of from 20,000 gauss to 100,000 gauss to much higher levels will effect total arrests of many diseases, such as all forms of cancers and other difficult complaints. We base this statement on new findings. These findings we believe will soon be released as fact within scientific releases coming from American and Puerto Rican researchers. Also, in this field of research will be releases of great international attention made by the Soviet Union's scientific research community. In this book we have stressed the importances of low to medium energies for the use of establishing the new concepts and laws of applied magneto magnetic physics necessary to understand this new and valuable science.

CONCLUSION

Man and all life forms are subject to a changing magnetic environment, and man should be educated to this fact at an early age. At the present this is not in the realm of primary instruction. Efforts should be forthcoming by responsible educators in the foreseeable future to bring about this better understanding of man and his environment. Nature plays the all-important role in man's existence on earth, yet we choose to manufacture artificial effects that override the balanced plans of nature.

Magnetism is a natural science. It is also a natural energy. This energy is generated by the nature of things, basically the atom itself. The earth is in fact a giant magnet. The energies provided affect all living systems on the face of the earth.

When we act to harness this form of natural energy, in making magnets that are artificial in that sense, we are duplicating nature's work by developing nature's natural energies. Then we discover that the basic and far-reaching approach in many fields, if not all fields of science, including the science of the medical arts, can be improved remarkably, by turning away from chemical reactions to a more normal and far more efficient use of controlled and proper directed magneto magnetic energies. Therefore, we hope this book will challenge the youth and the physicists of today, the scientific community as exploring scientists, to explore this new and exciting scientific probe, with a new outlook and new approach for a better world, the world of tomorrow.

ABOUT THE AUTHORS

ALBERT ROY DAVIS

Scientist, born in Halifax, Nova Scotia, Canada, June 18, 1915; parents, William Albert and Annie Agnes (Robinson), England. Came to United States, 1936; naturalized American citizen, 1936; attended University of Florida, 1936. Owner and manager, Albert Roy Davis Research Laboratory, Green Cove Springs, Florida, 1938—;writer technical papers radiological fallout, AEC, 1945-1946. Associate professor, biomagnetic sciences, Naihati (West Bengal, India) Research Center, 1964-1968; consultant to the board, 1965—; recipient of a number of honorary doctor degrees in science; Director of the Albert Roy Davis Aerial Phenomena Research Association and the United Science Federation, Green Cove Springs, Florida. Served under contract Air Transport Command, USAAF, Port Security division USCG, 1942-1943. Acknowledgment for work introducing biomagnetic sciences to scientists and doctors in India by Prime Minister, 1971; acknowledgment by industrialists in Japan on work to serve humanity through science of biomagnetics; author of many technical and scientific manuscripts on applied sciences, also over 370 general science courses used and adapted for grade schools, high schools and colleges in the United States and many nations of the world. Inventor. Developed methods of stimulation in milk products by application of magnetic fields and energies, 1965; methods in ecology studies to control nitrogen in water by biomagnetic molecular magnetic stimulation, 1969. Address: Post Office Box 655, Green Cove Springs, Florida, 32043.

WALTER C. RAWLS, JR.

Scientist, lawyer, business executive, born in Richmond, Virginia, September 13, 1928; parents, Walter Cecil and Ella Townsend (Freeman). University of Missouri, degree in arts and sciences, 1951; Juris Doctor degree, Washington University, St. Louis, Missouri, 1958; appointed by the Chancellor to permanent advisory committee, Washington University Law School, 1970—; Member of the Jacksonville, Florida, and American Bar Associations, International Bar Association, International Platform Association, Arbitrator for American Arbitration Association; Community Leader of America award, 1968; Trial Attorney, 1958-1969; Director, Georgia-Florida Oil & Refining Company, Inc., F.I.D. International, BioMagnetics International, Inc., Funds, Inc., Canvi-Andor, International Film Corporation, British West Indies Capital Reserves Ltd.; author of manuscripts in history and applied sciences. Awarded Honorary degree, doctor of science, 1973; acknowledged in national and international directories; married, Sheila (Kirsch), Doncaster, Yorkshire, England; sons, Richard Wayne and James David. Address: Post Office Box 655, Green Cove Springs, Florida, 32043.

GENERAL RESEARCH REFERENCES

H. S. Alexander, *Amer. J. Med. Electr.*, 1 (1962), 18]

M. F. Barnothy, ed., *Biological Effects of Magnetic Fields* (2 vols.; New York: Plenum Press, 1964 and 1969)

G. M. Baule and R. McFee, *Am. Heart J.*, 55 (1963), 95

D. Cohen, *Science*, 175 (1972), 664

D. Cohen, E. A. Edelsack and J. Zimmerman, *Appl. Phys. Letters*, 16 (1970), 278

R. Damadian, *Science*, 171 (1971), 1151

A. d'Arsonval, *C. R. Soc. Biol.*, 48 (1896), 450

L. D. Davis, K. Pappajohn and I. Plavnieks, "Bibliography of the Biological Effects of Magnetic Fields," *Fed. Proc.*, 21, Sup. 12, Part II (Sept. 1962), 1-38

J. W. Devine and J. W. Devine, Jr., *Surgery*, 33 (1963), 4

J. Driller, W. Casarella, T. Asch and S. K. Hilal, *Med. & Biol. Eng.*, 8 (1970), 15

M. Eibschutz, *et al.*, *Nature*, 216 (1967), 1138

M. Equen, *Magnetic Removal of Foreign Bodies* (Springfield, Ill.: Charles C. Thomas, 1957)

M. W. Freeman, A. Arrot and J. H. L. Watson, "Magnetism in Medicine," *J. Appl. Phys.*, 31S (1960), 404

E. H. Frei, *et al.*, *J. Appl. Phys.* 39 (1968), 999

D. B. Geselowitz, *Biophys. J.*, 7 (1967), 1

W. Gilbert, *"De Magnete Magneticisque Corporibus et de Mango Tellure,"* *Physiol. Nova* (London: 1960)

W. Haberditzl, *Nature*, 213 (1967), 72

S. K. Hilal, W. J. Michelsen and J. Driller, *J. Appl. Phys.* 40 (1969), 1046

G. C. Kimball, *J. Bact.* (1938), 109

A. Kolin, "Evolution of Electromagnetic Blood Flowmeter," *UCLA Forum Med. Sci.* (1970)

A. Kolin, *Physics Today*, 21 (Nov. 1968), 39

M. M. Labes, *Nature*, 211 (1966), 968

F. T. Luborsky, B. J. Drummond and A. Q. Penta, *Amer. J. Roentgen*, 92 (1964), 1021

S. Maeshima, "Magnetic Healing Apparatus," U.S. Patent No. 1,421,516 (July 1922)

J. Magrou and P. Manigualt, *C. R. Acad. Sci.*, 233 (1946), 8

R. McFee and G. M. Baule, *Proc. IEEE* 60 (1972), 290

P. H. Meyers, F. Cronic and C. M. Nice, Jr., *Amer. J. Roentgen*, 90 (1963), 1068

T. Nakamura, *et al.*, *J. Appl. Phys.* 42 (1971), 1320

P. W. Neurath, *Biological Effects of Magnetic Fields* (New York: Plenum Press, 1969)

L. A. Pirusian, *et al.*, *IZV. Akad. Science SSSR Biol.*, S4 (1970), 535

A. S. Presman, *Electromagnetic Fields and Life* (New York: Plenum Press, 1970)

A. Rosen, G. T. Inouye and A. L. Morse, *J. Appl. Phys.*, 42 (1971), 3682

Symposium on Application of Magnetism in Bioengineering, *IEEE Trans. Magnetics*, MAG-6 (1970), 307-375

J. A. Taren and T. O. Babrielsen, *Science*, 168 (1970), 138